Uwe Harms (Ed.)

Supercomputer and Chemistry

IABG Workshop 1989

With 46 Figures and 24 Tables

Springer-Verlag
Berlin Heidelberg New York London Paris
Tokyo Hong Kong Barcelona

Uwe Harms
debis Systemhaus GmbH
LRZ Ottobrunn
Leibnizstraße 7, D-8012 Ottobrunn
Federal Republic of Germany

ISBN 3-540-52915-2 Springer-Verlag Berlin Heidelberg New York
ISBN 0-387-52915-2 Springer-Verlag New York Berlin Heidelberg

This work is subject to copyright. All rights are reserved, whether the whole or part of the material is concerned, specifically the rights of translation, reprinting, reuse of illustrations, recitation, broadcasting, reproduction on microfilms or in other ways, and storage in data banks. Duplication of this publication or parts thereof is only permitted under the provisions of the German Copyright Law of September 9, 1965, in its current version, and a copyright fee must always be paid. Violations fall under the prosecution act of the German Copyright Law.

© Springer-Verlag Berlin Heidelberg 1990
Printed in Germany

2151/3140-543210 – Printed on acid-free paper

Preface

Supercomputer and Chemistry is the name of a series of seminars, which the Industrieanlagen-Betriebsgesellschaft (IABG), Ottobrunn near Munich, started in 1987. This third meeting stressed the fields of computational science, supercomputing and computer-aided chemistry. Moreover, the current situation in the supercomputer market as a whole, particularly in Germany, and the trends to be expected were discussed.

The new generation of graphic workstations such as StARDENT have the power of minisupercomputers. Some performance results are presented and comparisons with other machines are made. One of the most exciting prospects for improving the performance of computers is parallel processing. Especially, transputers seem to give unlimited computing speed, in effect a Cray on your desk. We examine the technology of transputers and their usage in industrial and research projects. The user will have a formidable task in parallelizing software.

The second part of the seminar addressed the usage of mainframes and supercomputers in the chemical industry. The interplay of experiments and computer-aided drug design was highlighted by speakers from Sandoz, Boehringer-Ingelheim and Merck. There is still one open question when using numerical methods, i.e. whether all the relevant and important conformations have been obtained. Certainly the computational results have to be checked and verified against experimental results.

Furthermore, the benefits, disadvantages and the reduction in costs and time in using supercomputers in pharmaceutical research were discussed.

Finally, the third part was dedicated to a new method, density functional calculations. This theory has been successfully used to calculate the physical properties of solids. The applicability of this method for molecules was discussed. Compared with the known Hartree-Fock SCF-calculations, this theory should give faster and better results. Some of the problems, especially the application in industrial environments, were discussed at length.

Without the support of my two colleagues, Dr. R. Iffert and N. McCann, this third workshop and its program would not have been possible.

June 1990 U. Harms

Table of Contents

U. Harms, IABG, Ottobrunn
Supercomputers - The Situation Today — 1

W. Gentzsch, FH Regensburg
Graphic Supercomputing on the Ardent and Stellar
Graphics Supercomputers — 11

G. Häußler, M. Grüner, CESYS, Erlangen
Transputer - A General Survey — 27

S. Streitz, PARACOM, Aachen
Transputers in Technical Applications — 45

U. Wedig, A. Burkhardt, H.G. v. Schnering, MPI, Stuttgart
Quantum Chemical Calculations on MIMD Type Computers — 51

U. Meier, Universität Bochum, R. Vogelsang, SUPRENUM, Bonn
Parallelizing an SCF Program on SUPRENUM — 69

H.P. Weber, SANDOZ, Basel
Drug Design: A Combination of Experiment and
Computational Chemistry — 89

G. Barnickel, MERCK, Darmstadt
Conformational Analysis of Peptides Using Molecular
Dynamics — 91

H. Köppen, Boehringer Ingelheim
The Use of Supercomputers in Medicinal Chemistry
Examples from Peptide and Protein Projects — 99

B. Delley, Paul Scherrer Institut, Zürich
Local Density Functional Calculations on Properties
of Large Molecules — 115

R.O. Jones, D. Hohl, KFA Forschungszentrum Jülich
Density Functional Calculations with Simulated Annealing
- New Perspectives for Molecular Calculations — 127

Authors' Index — 143

Supercomputers – The Situation Today

U. Harms
Industrieanlagen-Betriebsgesellschaft, Einsteinstr. 20, D-8012 Ottobrunn

Abstract: The current situation in the super- and minisupercomputer market is changing rapidly. New manufacturers try to place high-performance computers in the market and well known vendors stop their activities in supercomputing. The performance of these new supercomputers and the distribution of the machines by countries and institutions is dicussed. A list of the supercomputers installed in Western Germany is included.

1. SUPERCOMPUTER INDUSTRY

In 1989 the situation changed dramatically. One of the manufactures with the longest experience in supercomputing and technical and scientific computing cancelled its activities. ETA-systems, a CDC-subsidiary, stopped the production of its ETA-series of supercomputers because of marketing and financial reasons. In Germany, four ETA-systems had been sold and others waiting for signature (DWD, Offenbach, University of Hannover, RWTH Aachen, University of Colougne).

In the meantime, the list of manufactures and computers, which fail to meet the market in the last two years, is becoming longer and longer:

ETA with ETA 10	Supercomputer
MUTIFLOW TRACE	Minisupercomputer
SCS with SCS 40	Minisupercomputer
CYDROME/PRIME	Minisupercomputer
GOULD with NP series	Minisupercomputer
EVANS & SUTHERLAND with ES 1	parallel processing system
FPS with T-series	parallel Transputer system.

It can therefore be a problem for a potential industrial user to choose a system that will be supported for several years. Sometimes, a new system is announced, as in the case of the ES1, and a few months later the project is cancelled because of the hard- und software expenses during the development phase.

U. Harms (Ed.)
Supercomputer and Chemistry
© Springer-Verlag Berlin Heidelberg 1990

2. NEW SUPERCOMPUTER SYSTEMS

In 1989 several new supercomputers or series have been announced. CRAY announced and delivered the 1-8 processor system CRAY Y-MP. Siemens announced its (Fujitsu's) VP 2000-series designated S-serie from S 100/10 to S 600/20. The special feature of these computers is that a second scalar unit is available as an option. Having a job with a lower degree of vectorization, the operating system can switch to the other scalar unit and use the vector unit when it is idle. So the throughput of the whole system can be optimized.

NEC is the first Japanese manufacturer to announce a really parallel supercomputer. The NEC SX-3-series has a maximum of four vector processors. First deliveries will be in the third quarter of 1990. There are certainly open questions concerning the parallelizing software tools, compilers, operating system, etc.

The German supercomputer project SUPRENUM delivered its first machine. The first system with four clusters (= 32 processors) was installed at the GMD (Gesellschaft für Mathematik und Datenverarbeitung, Society for mathematics and informatics) at the end of 1989. It will be expanded to the maximum of 32 Clusters (= 256 processors) in the spring 1990. The first reports on experiences in the usage and user friendlyness of this machine will soon be available. Some experiences with the prototype will be discussed in the paper by U. Meier and R. Vogelsang, included in these proceedings.

In 1990 another supercomputer project in Karlsruhe Germany, from iP-systems will offer a massively parallel system with a tree structure. The architecture itself is proven but the system is not yet available with high-speed chips.

The well kwown Hypercube-architecture is supported by INTEL, in the IPSC-series with the well known 80386/87 processors. That system with 128 processors has a supercomputer power. Now INTEL is developing a prototype system for DARPA with 2000 processors using the i860 microprocessor. The main memory will be expanded to 128 Gigabytes, the computational speed is expected to be 128 Gigaflops. The prototype will be available at the end of 1991. A small series starting from 8, ending with 128 processors is now available. The computational power and the memory size gives a good alternative to the big supercomputers. Surely, a very interesting new system; all the now known communication problems must be solved by that date.

Computer	Clock cycle (Nanosec)	Vector-Processors	Memory (MByte)	Performance (MFLOPS)
CRAY Y/MP1	6	1	256	330
CRAY Y/MP8	6	8	1024	2700
Hitachi S820/20	4	1	128	375
Hitachi S820/80	4	1	512	3000
NEC SX-3/11	2.9	1	512	1370
NEC SX-3/44	2.9	4	2048	22000
Siemens S100/10	4	1	1024	500
Siemens S600/10	4	1	2048	4000
Suprenum (16)	100	1 (16)	128	320
Suprenum (256)	100	16 (256)	2048	5120
iPSC/2-d7	60	128	1024	840
iPSC/860-8	25	8	64	480
iPSC/860-128	25	128	2048	7600
CRAY 3	2	16	4096+	16000 (1990)
CRAY C90	4	16		16000 (1991)
CRAY 4	1	64	8000+	128000 (1992)
DEC 6000 VF	28	1-2	256	45/90
DEC 9000 VF	16	1-4	512	125/600
Alliant FX 2808	25	8	64	320
Alliant FX 2828	25	28	1024	1120

Table 1: Performance of new supercomputers and some hardware specifications; in parantheses, the year of customer delivery. The plus sign means that the main memory may be larger.

All the supercomputer people are still waiting for the Cray 3 built of gallium arsenide. The machine has been expected for several years and now it is hoped, that it will be available in 1990. There are two other machines, which are under development. The Cray C90 is the successor of the X-MP, Y-MP series and will be available in 1991. There will be a big increase in performance as compared with the Y-MP, a factor of three per processor.

The Cray 4 should be the successor of the Cray 3, but it could be, that it will be available earlier than the Cray 3.

A very important aspect of all the machines is the size of the main memory. The plus sign in table 1 indicates that the size might be larger when these computers are delivered.

The number two of the worldwide computer manufacturers , DEC, joined the supercomputing community. DEC now offers a vector facility, like IBM, for both its 6000-series for the minisupercomputer market and its 9000-series in the performance range of the IBM 3090 vector facility. The experts have been waiting for years for this architecture and technology. It will be interesting to see what market share DEC will be able to corner and how powerful the vectorization and parallelization, offered by the compiler, will be.

Alliant has announced its new series, the FX 2800, at the beginning of 1990. Now you can choose a system starting from eight and ending with 28 processors. The processors can be clustered, then these clusters can work on different jobs in parallel.

The FX 2800 series uses the new INTEL i860 chips, conservativly, Alliant estimates a maximum speed of 40 MFlops. The most interesting thing is the so called PAX (parallel architecture extended) standard. With this standard software a program will run on any INTEL i860 computer without any changes, that uses this standard. Here the parallelism is included.

Convex is planning a new series, perhaps the C3xx, the number of processors might be increased to 8, the processor speed by a factor of 3 or 4 (estimated). That means that a C3xx processor lies between the Cray 1 and Cray X-MP1 in its computing power.

3. SUPERCOMPUTERS WORLDWIDE

There are about 500 supercomputers installed worldwide. Most of the systems, about 40% is used in universities or research institutes. Especially in the USA, nearly 15% are used for defence applications. The supercomputers have had great success in aerodynamics and aerospace industries, which represent nearly 15% of the worldwide market. Another old application area is seismic exploration with 10% market share. Nowadays, environmental research and models are gaining rapidly representing about 5% of the supercomputers delivered. Nuclear energy have 5%, the automotive industry about 5%, the chemical and pharmaceutical industries and service bureaus 5% of the supercomputers in use.

The distribution of supercomputers by manufacturers is changing. As ETA is out of the supercomputer race, the Japanese manufacturers are expanding. The situation at the moment:

CRAY	56%
Fujitsu	21%
CDC/ETA	12%
NEC	5%
Hitachi	6%

Table 2: Distribution of supercomputers by manufactures

It is rather difficult to get exact information on the supercomputers installed in different countries, especially in Japan. Table 3 is a good estimate and underlines the great importance, supercomputing has in Japan.

USA	50%
Japan	25%
Europa	20%
* Great Britain	5%
* France	5%
* Germany	5%
other countries	5%

Table 3: Distribution of supercomputers by countries

3.1 SUPERCOMPUTING IN CHEMISTRY

The usage of these machines is rapidly growing in the chemical industries. The German industry is using minisupercomputers but is thinking of installing larger machines. Other countries have supercomputers in this field.

USA	
Du Pont	Cray X/MP 24
Monsanto	Cray X/MP EA 116
Ely Lilly & Co	Cray 2
Scripps Clinic	Cray X/MP EA 116
(agrochemistry, pharma, biotechnologies)	
Japan	
Sumitomo chemical engineering	Cray X/MP EA 116
unnamed companies in Japan	7 Fujitsu VP-systems

Table 4: Supercomputers in chemical industries

5. SUPERCOMPUTERS IN GERMANY

Since the first installation of a supercomputer in 1979, there has been rapid growth in the total number of systems and processors. Table 5 contains the list of institutions with the installation date of new machines. It is separated in three parts, the universities or university environment, research institutes and, finally, industry. The institutes are listed in the order of the installation date of their first supercomputers.

The Siemens supercomputers are built by Fujitsu, the S-series (S100, 200, 400, 600) is identical to the Fujitsu VP 2000-series.

1989/1990 brought an incredible improvement of computing power in Germany. Several universities, reseach institutes and industrial companies have doubled their capacity.

RRZN, University of Hannover, will be one of the first institutions that will install the first multiprocessor system built by Fujitsu. It will be very interesting to hear of the initial results in parallelization.

	Date	supercomputer	Date	new machine
Universities				
Bochum	1981	CDC Cyber 205		
Karlsruhe	1983	CDC Cyber 205	1988	Siemens VP400EX
			1990	Siemens S600
Stuttgart	1983	Cray 1M	1986	Cray 4 (4 Proz)
Berlin (ZIB)	1984	Cray 1M	1986	Cray X-MP 2
Kaiserslautern	1986	Siemens VP100		
Kiel	1987	Cray X-MP 1	1988	Cray X-MP 2
München (LRZ)	1988	Cray X-MP 2	1989	Cray Y-MP 4
Hannover (RRZN)	1990	Siemens VP200EX	1990	Siemens S400/10
		3. stage	1991	Siemens S400 (MP) multiprocessor system
Aachen (RWTH)	1990	Siemens VP200EX	1991	Siemens S400/10
			1992	Siemens S600/20 or S400 (MP)
Research institutes				
MPI Garching	1979	Cray 1	1986	Cray X-MP 2
MPI Hamburg	1985	CDC Cyber 205	1988	Cray 2 (4 Proz)
KfK Karlsruhe	1987	Siemens VP50	1990	Siemens VP400EX
KFA Jülich	1983	Cray X-MP 2	1989	Cray X-MP 4
HLRZ Jülich	1987	Cray X-MP 4	1989	Cray X-MP 8
DLR Oberpfaffenhofen	1983	Cray 1S	1987	Cray X-MP 2
		3. stage	1990	Cray Y-MP 2
GMD St. Augustin	1990	SUPRENUM (256)		
Industry				
Prakla-Seismos Hannover	1981	CDC Cyber 205	1989	Cray X-MP 1
IABG Ottobrunn	1985	Siemens VP200		
EDS (Opel) Rüsselsheim	1985	Cray 1S	1989	Cray X-MP 2
Siemens München	1985	Siemens VP200		
Siemens München	1988	Siemens VP100EX		
Gouvern. dep. Bonn	1985	Cray X-MP 4		
Daimler Benz Stuttgart	1987	Cray X-MP 2	1989	Cray Y-MP 2
Volkswagen Wolfsburg	1987	Cray X-MP 1	1989	Cray X-MP 2
DWD Offenbach	1987	CDC ETA-10	1990	Cray Y-MP 4
BMW München	1988	Cray X-MP 2		
Continental Gummi Hannover	1989	Cray X-MP		

Table 5: Supercomputers in Western Germany

Some of the institutions in table 5 are listed below:

ZIB: Konrad Zuse Zentrum for information technology, computer center for the universities of Berlin and other counties in northern Germany.

LRZ: Leibnizrechenzentrum, Bavarian Academy of Science, computer center for the Bavarian universities.

RRZN: Regionales Rechenzentrum für Niedersachsen at the University of Hannover, computer center for the universities of lower Saxony and counties in northern Germany.

RWTH: Rheinisch-Westfälische Technische Hochschule, computer center for the universities of Nordrhein-Westfalen.

MPI, Garching: Max Planck Society, institute of plasma physics.

MPI, Hamburg: Max Planck Society, computer center for climate research.

KfK: Kernforschungszentrum Karlsruhe, nuclear research center.

KFA: Kernforschungsanlage Jülich, nuclear research center.

HLRZ: Höchstleistungsrechenzentrum, central computer center for researchers in theoretical physics from all over Germany; it is supported by the KFA.

DLR: Deutsche Forschungs- und Versuchsanstalt für Luft- und Raumfahrt, computer center for the research in aerospace.

GMD: Gesellschaft für Mathematik und Datenverarbeitung, computer center for the society of mathematics and computer sciences.

Prakla-Seismos: seismic exploration.

IABG: Industrieanlagen-Betriebsgesellschaft, supercomputer services for industry.

EDS: subsidary of General Motors, in Germany, Opel automotive industry.

Siemens: VP 200 is used for circuit and chip design, VP 100 EX for benchmarking and vectorization of application software.

DWD: Deutscher Wetterdienst, computer center of the German Metereological Institut the weather forecasts.

There are some other projects in Germany that plan the installation of supercomputers. In the county of HESSE, there is a working group that is discussing the location and the type of machine for all the universities of this county.

In Colougne, there are plans for a supercomputer, but since Aachen which is in the same county has just been equipped, they will surely have to wait.

In Bavaria, there are plans for a second supercomputer at the University of Erlangen, but they are just in the discussion phase. As Erlangen is working in the Suprenum project; perhaps such a machine might be installed.

For the whole Germany there are some innovative research projects, one is in the hypersonic flow for aerospace projects. Research groups all over Germany need supercomputer power. At the moment they use different sources but perhaps there is a central solution.

Two years ago, G. Hohlneicher at the first IABG-workshop had a talk on the computational situation in the theoretical chemistry in Germany. Several important professors had written a memorandum that indicates: the theoretical chemistry needs a supercomputer. Perhaps some users will support these ideas, it then might be granted by the German ministry of technology and research, to form a supercomputer center for the chemistry similar to the HLRZ at Jülich.

References:
G. Hohlneicher: Vorstellung der theoretischen Chemiker zur Ausgestaltung eines Hochleistungsrechenzentrums für die theoretische Chemie, Supercomputer-Seminar 1987, IABG.

Graphic Supercomputing on the Ardent and Stellar Graphics Supercomputers

W. Gentzsch
FH Regensburg, Pruefeningerstr. 58, 8400 Regensburg, FRG

1. INTRODUCTION

Both the size of scientific and engineering problems and the quantity of data generated present a large challenge to researchers: namely, how to understand the results of their computations. Solving these problems interactively is essential and requires significant computation and graphics power. While graphics and computing hardware have made rapid strides in the last few years, the software available to researchers has not kept pace. This lack of software is underscored by the recent emergence of the graphics supercomputers, which combine minisupercomputer-class computational power with 3D graphics capabilities that support real-time, interactive display of scientific and engineering data.

ARDENT's TITAN and STELLAR's GS series both are graphics supercomputers which provide scientific computations and high-performance graphics. They are multiprocessors with up to four processing (or stream) units that run separately or in parallel on one task and consist each of integer and vector units and additional synchronization registers for parallel processing.

The vector units support basic vector operations, gather/scatter, merge, compress and mask, reduction and chained operations. They contain independent arithmetic functional units (an adder, multiplier, divider), and load/store pipes. Buses supply data for the vector unit at high bandwidth. Because of these vector and parallel properties, the user of an Ardent or Stellar system is faced with the same problems of software implementation as are other users of todays supercomputers and mini-supercomputers, see e.g. [3,4,5,6].

U. Harms (Ed.)
Supercomputer and Chemistry
© Springer-Verlag Berlin Heidelberg 1990

After some general remarks on visualization in chapter 2, a description of the hardware and software features including graphics software such as DORE and AVS follows in chapters 3 and 4.

2. VISUALIZATION

Visualization in scientific computing is emerging as a major computer-based field, with many problems and boundaries, a commonality of tools and terminology, and a cohort of trained personnel, [8]. It can bring enormous leverage to bear on scientific productivity and the potential for major scientific breakthroughs, at a level of influence comparable to that of supercomputers themselves. It can bring advanced methods into technologically intensive industries and safeguard the effectiveness of the scientific and engineering communities. Major advances in visualization and effective diffusion of its technologies will drive techniques for understanding how models evolve computationally, tightening design cycles, better integrating hardware and software tools, and standardizing user interfaces.

Visualization will also provide techniques for exploring an important class of computational science problems, relying on cognitive pattern recognition or human-involved decision making. New methods may include guiding simulations interactively and charting their parameter space graphically in real time. Significantly more complexity can be comprehended through visualization in scientific computing techniques than through classical ones.

Visualization is a method of computing. It transforms the symbolic into the geometric, enabling researchers to observe their simulations and computations. It enriches the process of scientific discovery and is revolutionizing the way scientists do science. It embraces both image understanding and image synthesis; that is, it is a tool both for interpreting image data fed into a computer and for generating images from complex multidimensional data sets. Visualization studies those mechanisms in humans and computers which allow them in concert to perceive, use, and communicate visual information. It unifies the largely independent but converging fields of

* Computer graphics
* Image processing

* Computer-aided design
* Signal processing
* User interface studies

Also, scientists not only want to analyze data that results from supercomputations, they want to interpret what is happening to the data during supercomputations. Scientists want to steer calculations in close-to-real-time; they want to be able to change parameters, resolution, or representation, and see the effects. Scientists want to be able to interact with their data.

The most common mode of visualization today is batch: compute, generate images and plots, and then record on paper, videotape, or film. Interactive computing would be an invaluable aid during the scientific discovery process, as well as a useful tool for gaining insight into scientific anomalies or computational errors.

Scientists need an alternative to numbers. A cognitive possibility and technical reality is the use of images. The ability of scientists to visualize complex computations and simulations is absolutely essential to ensure the integrity of the analysis, to provoke insights, and to communicate about with others.

There are several visually oriented computer-based technologies in existence today. Some have been exploited by the private sector, and off-the-shelf hardware and software can be purchased, as with the graphics supercomputers of Ardent and Stellar. Others require new developments; some point to new research areas. Visualization technology, well integrated into today's workstation, has found practical application in such areas as product design, electronic publishing, media production, and manufacturing automation.

Traditionally, scientific problems that required large-scale computing resources needed all the available computational power to perform the analysis or simulation itself. The ability to visualize results or guide the calculations requires substantially more computing power.

To date visualization software tools readily available to the computational scientist and engineer have fallen into two categories: graphics subroutine libraries and animation applications. Graphics libraries such as PHIGS+, GL, GKS, and SIGGRAPH CORE are examples of low-level collections of graphics operations. These libraries represent the traditional, structured language approach to programming. However, they fail to hide the basic complexity of the visualization problem, since their effective use requires programmers to understand the graphics primitives and data structures inherent in the library.

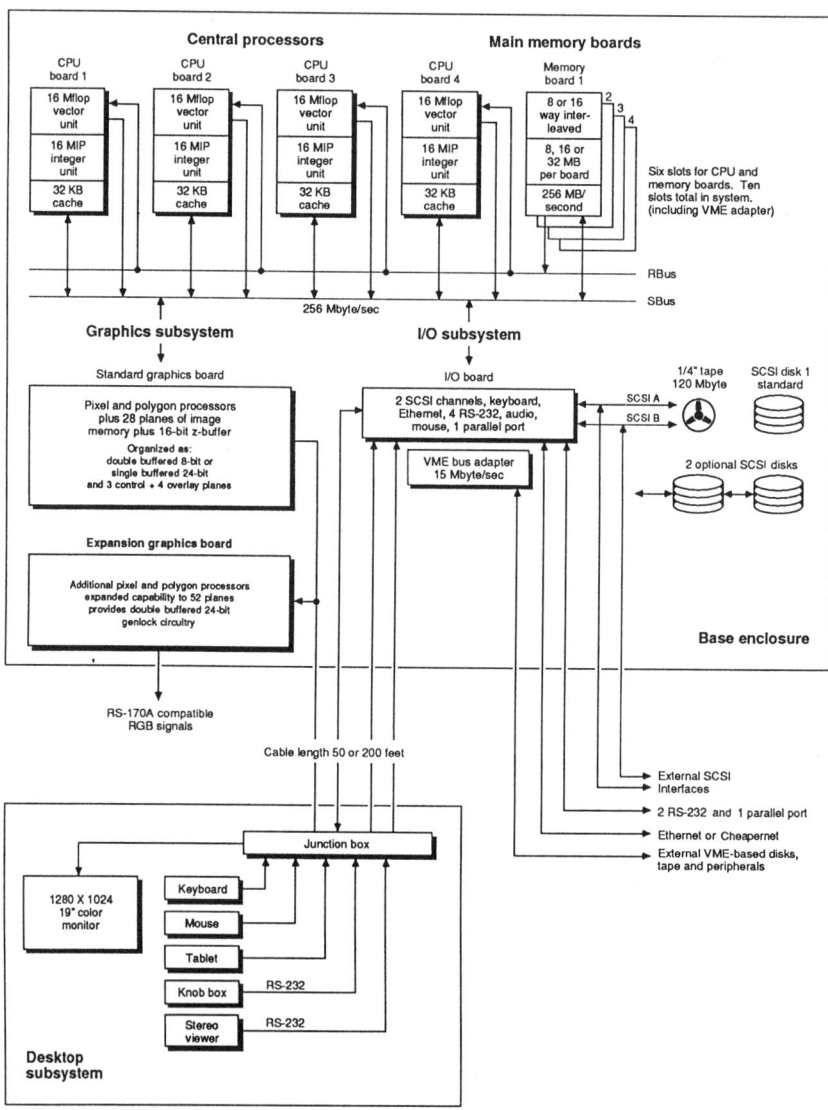

Figure 1: Architectural Overview of the ARDENT Titan.

Animation packages such as MOVIE.BYU and products from Wavefront and Alias are some of the more common visual applications currently used by scientists. These programs are used in a postprocessing mode after a numerical simulation is run on a supercomputer or mini-supercomputer, typically in batch mode, and after the data have been transferred to the display system. The traditional drawback of these packages is that they frequently are too restrictive in their capabilities: Either they are tailored for such specific disciplines as computational chemistry or fluid dynamics or their data formats are limited to such geometric primitives as points, lines, and polygons.

3. ARDENT'S TITAN

Titan is a symmetric multiprocessor with up to four processing units that run in parallel, see Figure 1. Each processor consists of two distinct parts: an integer processor and a vector processor. The integer processor is an off-the-shelf 32-bit reduced-instruction-set-computer (RISC) microprocessor from MIPS Computer Systems for high clock rate and single-cycle execution. Two direct-mapped caches for non-floating-point data and for instructions provide fetch-ahead and cycle independently.

Titan's 64-bit vector unit is basically Cray-like in its organization, with its proprietary vector processor based on the WTL 2264/5 floating-point chip set from Weitek Corp. Each vector unit has a peak floating-point execution rate of 16 Mflops.

Like Cray the vector unit supports scatter/gather, hardware merge, compress and mask, reduction, and chained operations. Each unit contains a high-bandwidth vector register file that contains 8,192 64-bit double-precision floating-point elements and can be software-configured into multiple vector registers of any length. Instructions execute much faster with data in registers before computation than when loading the variables into registers from memory. The vector register file also can address individual elements within a vector. And it can start and end a vector operation at any segment in the vector. Most other architectures can only address the first element of a vector. The register file is also dynamically reconfigurable, so the variables for several tasks can be maintained in the file at the same time.

In addition to incorporating the large, flexible register file, the vector processor makes heavy use of pipelining. The processor has six elements, which can be run in parallel: three independent, arithmetic functional units multiplier, and divider); and independent memory pipes (two for loading data from memory and one for storing results back to memory); All six elements can execute instructions simultaneously if the operation of one is not dependent on the outcome of another.

Titan's synchronous split-bus architecture allows the system to transfer 128 bits of data every clock cycle, thereby affording a sustained bus transfer rate of 256 Mbytes/s. One branch of the split bus is devoted to loading the vector unit. The other branch is used for all the bidirectional transfers betweeen memory and the vector unit, integer unit, graphics subsystem, and input and output ports of the system. The vector and integer units each have separate paths to the bus so they can run asynchronously but the integer unit is tightly coupled to the vector unit.

The integer unit connects to a separate instruction and data cache. For writing, the integer unit has a 4-word-deep write pipeline that buffers write operations from the processor. The processor puts a write operation in the pipeline and continues processing while the buffer writes the data to memory.

A separate bus watcher guarantees that if memory is changed and there is a value in cache which was affected by the change, the cache is invalidated. This removes from the compilers and operating system the chore of monitoring the write-through cache to see that it contains the most up-to-date data, for more details, see [1].

There are three opportunities for the compiler to improve processor performance. It can compile code to execute in parallel on the four processors in the system. It can change integer operations into vector operations. Or it can assign tasks to the integer processor in order to offload the vector processor. For example, the integer processor can prepare the next vector operation, while the current vector operation is still being completed.

To display data processed by the four processor boards, Titan comes with a high-performance graphics subsystem, see Figure 2. The vector processor performs graphics functions such as 4-by-4-matrix transformations. The fast RISC integer processor performs graphics operations requiring integer operations such as display-list processing. In addition, the display-list processing itself can be further parallelized to run on all four RISC processors at once. The vector processor also formats and passes commands to the pixel and polygon processors. After the image has been processed by the vector and integer processors, it is sent to proprietary pixel and polygon processors which have been dedicated to performing the rendering, polygon fill, translation, and other graphics operations. The graphics subsystem contains eight parallel pixel processors, two expandable polygon processors, and 24 image planes that can be organized as a double-buffered 8-bit frame buffer. It also has a 16-bit Z-buffer with four overlay planes and three control planes.

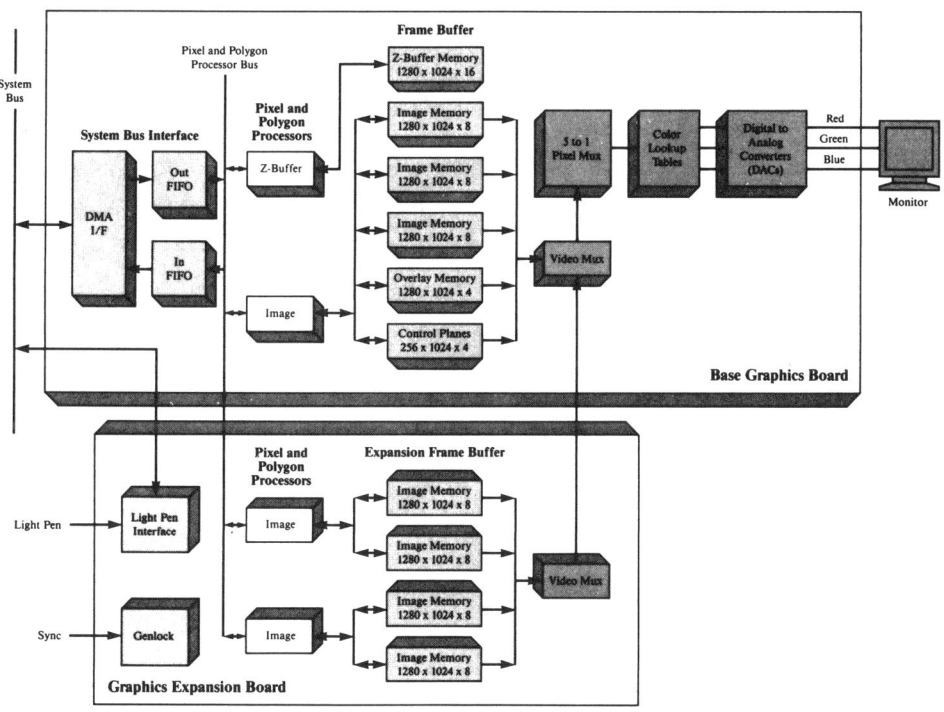

Figure 2: Titan's Graphics Subsystem.

An expansion board adds another 32 color planes that provide 48 image planes for double-buffered 24-bit full color. To produce a full-color image requires the 24 bits of data to describe every pixel on the screen. The expansion board also contains four additional pixel processors and another polygon processor. A fully configured system consists of 12 parallel pixel processors, three polygon processors, 48 image planes for double-buffered, 24-bit, fullcolor images, and a 16-bit Z-buffer with four overlay planes and three control planes.

Titan's peak hardware Gouraud shading rate is 50 million pixels per second; its software Phong shading rate is approximately 250.000 pixels per second per processor. The peak hardware vector drawing rate is 11.6 million pixels per second; its software antialiased vector drawing rate is 250.000 pixels per second per processor.

For interactive visulalization, Titan provides a high-level toolkit that takes particular advantage of the system's closely coupled architecture. This high-level package is known as the Dynamic Object-Rendering Environment, or Dore, [2]. It allows users to visualize the results of compute-intensive problems. It provides capabilities comparable to existing graphics libraries, such as Silicon Graphics GL, HP Starbase, Apollo GMR, and PHIGS/PHIGS+, as well as advanced features found in camera quality rendering packages, such as those from Wavefront and Pixar. Dore provides a hierarchical scene database with complete editing facilities. It also provides immediate mode graphics if the application prefers to maintain its own scene database.

Dore simplifies the task of attaching graphics to an application. It is a comprehensive library of graphics subroutines that lets developers create a wide range of integrated interactive and static full-color, three-dimensional imagery - without graphics-specific code. Developers build visual images of computational results by drawing on the Dore database of graphics primitives, appearance attributes, and geometric attributes, among other elements. Once an object has been described, the user "asks" the Dore integral renderer to produce an image by choosing interactively among different representations (points, wireframe, faceted, smooth-surface, or any combination) different shading models (ambient, diffuse, specular, reflections, and shadows), and different levels of realism.

In addition, there are several advanced features in Dore: a spatial-subdivision ray-tracer, a geometric compiler, flexible shading algorithms, advanced primitives and attributes like cubic solids, NURBs, polygonal meshes, transparency, solid-texture mapping, and more. Dore is also extensible, so developers and users can add their own attributes, primitives, textures, shading, and rendering functions with their own C or FORTRAN code.

When using Dore software to visualize the results of computation, users describe a scene, produce images with high information content from the scene database, and manipulate that data interactively and dynamically. The Dore toolkit's hierarchical scene database allows developers to create abstract scenes using five classes of objects:

PRIMITIVES in the Dore toolkit range from simple entities like points, curves, and text to more complex geometric objects like polygonal meshes, cubic solids, and non-uniform rational B-spline surfaces. Primitives can include not only geometry but also user-supplied data for illustrating the results of analysis.
APPEARENCE ATTRIBUTES define the visual properties of surfaces. They range from simple specifications like color and specularity to more complex properties like transparency, solid texture mapping, and environmental reflection. Appearance attributes also include tesselation parameters and geometric representations. Point, wireframe, and flat and smooth surface representations are supported.
GEOMETRIC ATTRIBUTES modify the overall geometry of primitive objects through such operations as translation, rotation and scaling.
STUDIO OBJECTS allow users to set up a virtual photography studio to produce images from a scene. This class of objects includes cameras and lights.
ORGANIZATIONAL OBJECTS allow users to collect objects into a hierarchical database and to call back the application to provide scene objects during execution.

4. STELLAR'S GS SERIES

The Stellar Graphics Supercomputer combines a large number of powerful functional units into a tightly-intergrated structure. The system includes several types of high-performance processors, see Figure 3:

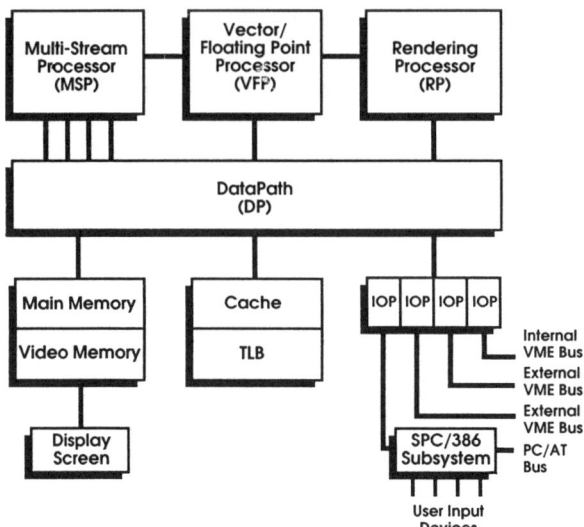

Figure 3: STELLAR Graphics Supercomputer Architecture.

The Multi-Stream Processor (MSP) can execute four instruction streams concurrently, and has special "concurrency registers" that enable fast inter-stream synchronization. The four streams share a single main memory, cache, and translation lookaside buffer, minimizing overhead when the streams are cooperatively executing a single program and operating on shared data.

The Vector/Floating Point Processor (VFP) contains four floating-point compute engines, configured to work separately or in tandem.

The Rendering Processor (RP) performs special graphics computations, such as per-pixel arithmetic for hidden-surface elimination ("Z-buffering"), depth-cueing, and shading. It is a microcoded engine that performs all pixel-level calculations. This unit has a Stellar-proprietary 16-processor SIMD architecture that runs at 320 million operations per second. Once it has been started with special-purpose vector graphics instructions, this processor runs independently of the MSP and VFP, providing the machine architecture with yet another level of concurrent processing. The Rendering Processor creates images of arbitary size in virtual memory. Visible portions are subsequently transferred on a demand basis to video memory, under control of the window system.

The Video Generation hardware scans out the image stored in video memory, maps pixels through a pseudo-color or true-color lookup table, and displays them on the high resolution (1280 x 1024 x 16/32 plane) color monitor.

Up to four I/O Processors (IOPs) handle communication with external busses, controllers, and peripheral devices. The SPC/386 Subsystem with its Intel 80386 processor acts as the system service processor and as a platform for MS-DOS applications.

These processors are synchronized by the system clock, which has a cycle of 50 nanoseconds.

The memory subsystem services all processors. A main memory of 16-128 MB can deliver 64 bytes to/from register or the Rendering Processor every 200 nanoseconds (320 MB/sec). The large cache (1MB) can deliver 64 bytes to/from registers every 50 nanoseconds (1.28 GB/second).

Data is routed among the systems processors through a central "data switchboard", which employs a DataPath architecture. The DataPath contains all the registers used by the machine's processors to perform integer, floating point, vector, graphics, and I/O operations.

The GS 1000's processors can operate simultaneously and independently. Moreover, they can operate cooperatively on a single application. For instance, a complex graphics application includes a tremendous amount of data movement and many different kinds of computation. Typically, an object is decomposed for purposes of image rendering into many triangles. At a particular instant, several kinds of work might be taking place:

* Multi-Stream Processor: Loading a vector register with the 3-D coordinate data that defines triangles.
* Vector/Floating Point Processor: Applying matrix transformation to coordinate data, loaded previously, that defines other triangles.
* Rendering Processor: Rendering a Gouraud-shaded image of still other triangles, whose coordinates have previously been loaded and transformed.

The Stellar Graphics Computer supports the implementation of high performance 3-D graphics applications in a multi-window environment:

Stellar PHIGS+ is an implementation of the ANSI/ISO PHIGS programming library, with the addition of PHIGS+ extensions. PHIGS enables FORTRAN and C programs to define, dynamically edit, and display hierarchical graphics databases.

Stellar PHIGS+ extensions include facilities for lighting, shading, and depth-cueing. Also included are additional graphics primitives, such as meshes and non-uniform rational B-spline (NURBS) curves and surfaces.

The Stellar X Window System is an implementation of Version 11 of the X Window System specification. This system allows multiple applications to run in a windowed environment. In particular, multiple Stellar PHIGS+ 3-D graphics applications can be active at the same time, each in its own window(s).

XFDI is a low-level programming library, through which Stellar PHIGS+ and Stellar X Window Sysem applications access the 3-D dynamic shaded graphics capabilities of the Stellar Graphics Supercomputer hardware. Programs can also make XFDI calls directly to access the maximum performance of the hardware, without invoking the higher-level functions of the Stellar PHIGS+ library.

The library routines make use of the Vector/Floating Point Processor, employing standard vector instructions (e.g. matrix multiplication), as well as graphics-specific machine instructions, e.g. clip checking.

Software and hardware components work together to implement a graphics pipeline - a series of operations that takes a simple graphical object from its logical 2-D or 3-D representation to its visual representation as displayed pixels:

* Primitives are generated by traversing a software-maintained (hierarchical) display list or directly by an application program fro its own database ("immediate mode").

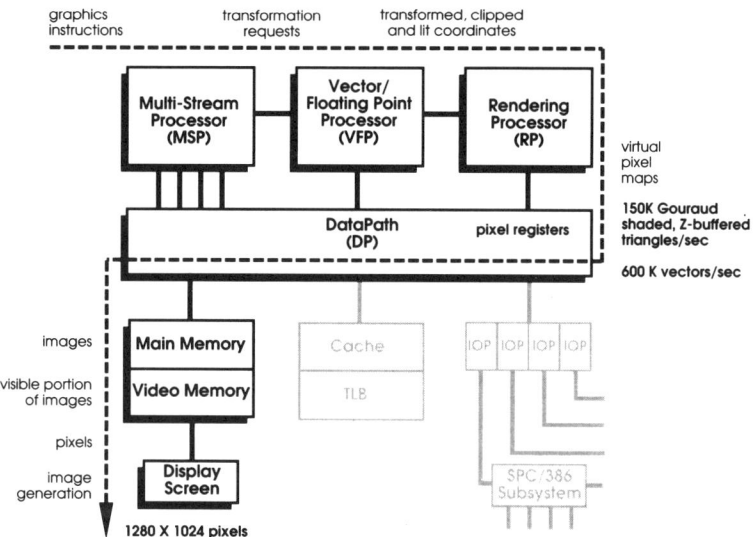

Figure 4: The Graphics Transformation Pipeline.

* These primitives are transformed, clipped, and lit by vector and graphics instructions executed by the MSP streams and the VFP.

* Transformed and clipped primitives are rendered according to a variety of lighting and shading methods by the Rendering Processor. The images are stored in arbitrary-sized virtual pixel maps. For 3-D primitives, the corresponding Z-buffers (and any additional buffers) are also stored in virtual pixel maps.

* Under control of the window system visible portions of virtual pixel maps are copied from virtual memory to displayable video memory.

* The Video Generation hardware scans out the image in video memory, maps pixels through the pseudo-color or true-color lookup table, and displays them on the high resolution 1280 x 1024 color monitor.

* Application programs can specify true-color, pseudo-color, or pseudo-color/stereo on a per-window basis. Each pseudo-color window can use a different color lookup table. True-color stereo viewing is supported on a full-screen basis.

Several innovative elements of the Stellar graphics pipeline, see Figure 4, provide unprecedented performance and flexibility:

Flexible Transformation Pipeline: Instead of implementing the entire graphics pipeline with special-purpose hardware, most graphics-related calculations are implemented in software, using standard arithmetic instructions and special graphics instructions. These calculations make full use of the machine's high-speed floating point and vector capabilities. Only pixel-oriented rendering operations are executed by the special-purpose Rendering Processor.

For example, primitives are transformed by multiplying their vertices (expressed as 4x1 homogeneous-coordinate vectors) by a 4x4 transformation matrix, using vector instructions. There are also vector instructions to perform clipping and lighting calculations.

This design exploits the commonality between graphics computation and general-purpose numeric computation, applying the supercomputer performance of the VFP to both types of processing. It also provides a flexible software implementation, while minimizing the amount of costly, dedicated hardware.

Virtual Pixel Maps: Images are generated not into video memory directly, but into virtual memory. This allows images and associated buffers, such as depth-buffers (Z-buffers), to be practically unlimited in number, size, and depth.

With Stellar's virtual pixel maps rendering technique, an image of the desired size is generated into virtual memory, and only visible portions are loaded as needed into video memory. The programmer can choose the depth of the Z-buffer, add other buffers (e.g. a pixel coverage Alpha buffer for anti-aliasing) or implement various forms of post-processing, compositing, image processing operations on images.

Instead of providing twice the amount of expensive video memory for double buffering of a fixed-size image, the Stellar provides n-way buffering of arbitrary-sized images. Such n-way buffering is very useful in providing smooth animation because rendering can take place in parallel with refreshing from a static image in video memory.

AVS (Stellar's Application Visualization System) is a framework that can be used to develop scientific visualization applications based on a model that integrates interactive visualization into the research and engineering process. It is targeted at scientists and engineers, rather than at software developers. The design goals of the system include the following:

* Ease of use. By employing direct-manipulation user interfaces (like those found in Apple Macintosh software) and simplifying the programmer's task, the system is more accessible to scientists and engineers. By exploiting the commonality among visualization applications, application development that takes weeks of tedious programming with current software may often be reducible to only a few hours of direct manipulation with AVS.
* Low cost. In contrast to large, monolithic third-party application software systems - which are expensive to develop and port to different hardware platforms - the goal is a framework that can integrate smaller-scale software components, which are less costly to develop.
* Completeness. Other visualization software focuses primarily on graphics rendering and viewing manipulations. AVS is designed to include the entire visualization process encompassing data input and transformation, as well as rendering.
* Extensibility. The approach assumes that scientists and engineers will need to extend and customize the software for their individual needs by adding their own algorithms and tailoring those provided in existing modules within the context of the visualization application framework.
* Portability. To be truly useful, the system must be available on the many platforms in heterogeneous computing environments. To foster portability, AVS is based on standards in areas like graphics libraries, windowing systems, operating systems, and languages.

In August 1989, Ardent Computer Corp. and Stellar Computer Inc., announced they plan to merge within the next months. Stardent Inc. will be created as a result of the merger. The companies plan to combine their product lines, with full compatibility achieved by early 1991. At the solftware level the respective products, which use UNIX V.3 and other industry standards, are already compatible. In the next 18 months they will be working toward having a completely unified product family with a single instruction set and full compatibility at the software source levels. Software developed on Ardent and Stellar machines will be portable to the new Stardent systems.

REFERENCES

1. Blank,S.G.: The Graphics Supercomputer: A Synergy of Advanced Scientific Computation and High-Performance Graphics. Super-Computing Summer 1988, pp. 5-9.

2. DORE Technical Overview. Ardent Computer Corp. CA 1988.

3. Gentzsch,W.: Die 2. Generation der Mini-Supercomputer. PIK 11 (1988) 3, pp. 171-174.

4. W.Gentzsch: Vergleich der Mini-Supercomputer Alliant FX/8, Convex C1-XP, FPS M64, Gould NP1, Multiflow TRACE 7/200 und SCS-40. PIK - Praxis der Informationsverarbeitung und Kommunikation. Carl Hanser Verlag, 11 (1988) 1 30 - 39.

5. W.Gentzsch, K.Neves, H.Yoshihara: Computational Fluid Dynamics: Algorithms and Supercomputers. Agardograph 311, 1988.

6. Gentzsch,W.:Benchmark Results for Shared-Memory Systems. Procs. 4th Int. Conf. Supercomputing, Santa Clara 1989.

7. Hockney,R.W., C.R.Jesshope: Parallel Computers. A.Hilger 1981.

8. McCormick, B.H., DeFanti,T.A., Brown,M.D.: Visualization in Scientific Computing. IEEE Computer Graphics and Applications, July 1987, pp. 61-70.

9. Stellar Graphics Supercomputer Modes GS1000. System Overview. Stellar Computer Inc. Newton MA. 1988

10. Upson, C.,u.a.: The Application Visualization System: A Computational Environment for Scientific Visualization. IEEE Computer Graphics and Applications, July 1989, pp. 31-42.

Transputer – A General Survey

Gerd Häußler, Markus Grüner
CESYS, Gesellschaft für angewandte Mikroelektronik mbH, D-8520 Erlangen

INTRODUCTION

Transputer - one more exotic appearance within the nearly immeasurable range of microprocessors? You might ask such a question, faced with the multitude of different processor types. The following article will point out the Transputer's unique position amongst microprocessors. Not just a few look at the Transputer as the European answer to new and old processor architectures pouring in mostly from USA and Japan. This is especially true since SGS-Thompson, which is an important leader in European technologies, has taken over control in Transputers and has entered the processor market with stimulating offensiveness.

INTERNAL ARCHITECTURE

"Transputer" actually is a generic term describing microprocessors that incorporate special provisions for coupling processors in a multiprocessornetwork without the need of connecting them to a common bus. This is similar, for example, to the term "signal processor" used in reference to DSPs. Therefore the term Transputer has not been protected by any copyright. Transputers of INMOS, Inc. are distinguished from other microprocessors especially in the following features:

- CPU combines RISC and CISC technologies
- 2K bytes or 4K bytes highspeed on-chip RAM
- variable length instruction codes
- serial communication channels known as links
- hardware process scheduler
- 64-Bit Floating Point Unit provided in the T800

integrated memory port
on-chip PLL for internal clock generation

Transputers use the latest VLSI (= very large scale integration) technology providing free programmable microprocessors with internal word widths of 16 bit for the M212, the T212 and the T222, and 32 bit for the T414, the T425 and the T800. The block diagrams of figure 1 illustrate the principal internal structure of the T414 and the T800 Transputers.

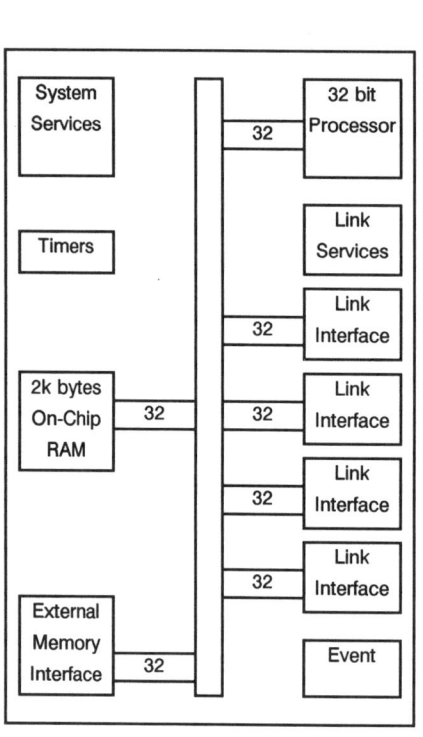 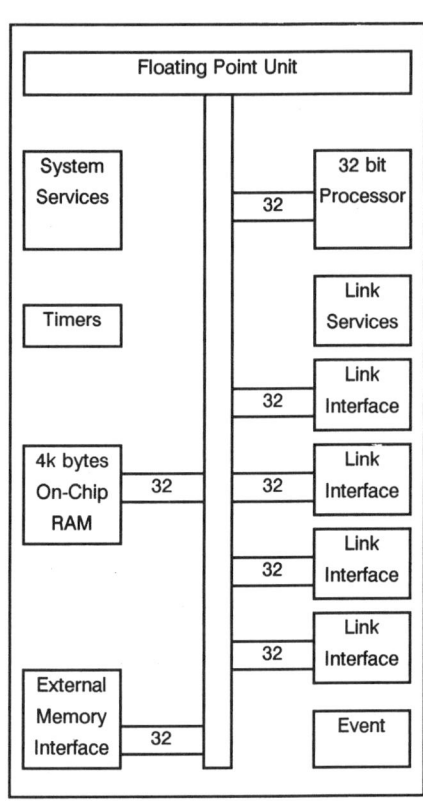

T 414   T 800

Figure 1: IMS T414 and T800 block diagram

Internally a Transputer consists of 2K bytes or 4K bytes of static RAM, the CPU, the communication ports, an interface to external RAM and, in the case of the T800, an FPU. All these basic elements claim approximately equal portions of the chip area and are coupled via a 32 bit wide bus (16 bits for M212, T212, T222). The block refered to as System Services contains the logic circuitry for initialization and control of the entire chip as well as for error handling and tools for chip analysis. The Event block is roughly comparable to the interrupts of other processors: through an event input an external device can request communication with the Transputer. Finally two internal timers are provided, one being incremented each microsecond the other with a resolution of 64 microseconds.

On-Chip RAM

Unlike many other typical RISC processors that need an increasing number of CPU registers to speed up memory access, the Transputer's CPU has just a few registers (8 in total). Among these, the "Workspace Pointer", provides relative addressing, i.e. relative to its current position. By directing this pointer to the on-chip RAM therefore, the entire RAM can effectively be used as register space (1024 registers, each 32 bits wide for the T800). The Transputer is however, flexible enough to allow small programs or especially time critical routines to coexist in the RAM. The on-chip RAM normally overlaps external memory but can be optionally inhibited by an external signal to the processor.

The Registers

Among the registers of the Transputer CPU the three known as A, B and C are arranged as a stack for address and integer computations. When a value is loaded into the CPU, the contents of the B register are first transferred to C, the contents of A are moved to B and then the value is loaded into register A. The procedure is reversed for storing a value; the contents of register A go to memory, B moves to A and C moves to B. The T800's FPU also contains the three registers AF, BF and CF, organized in the same manner.

Workspace pointer, instruction pointer and operand register are three more registers of the Transputer CPU. The workspace pointer, mentioned above, tracks the address of the local variables a process is using. The instruction pointer contains the address of the next instruction to execute and the operand register is used to set up complex instructions. Arithmetic expressions are evaluated in registers A, B and C. All binary operations, i.e. operations with two operands (e.g. add, multiply,...), operate on registers A and B. An addition, for example, adds registers A and B, stores the result in A and transfers the contents of register C to B. The original contents of registers A and B are lost. Non commutative operations are always performed as:

    A - Register = B - Register [Operation] A - Register

Since there is no hardware provided on a Transputer to recognize stack overflow, the programmer is completely responsible to ensure, that expression evaluation won't exceed the maximum stack depth of three, usually by using temporary variables.

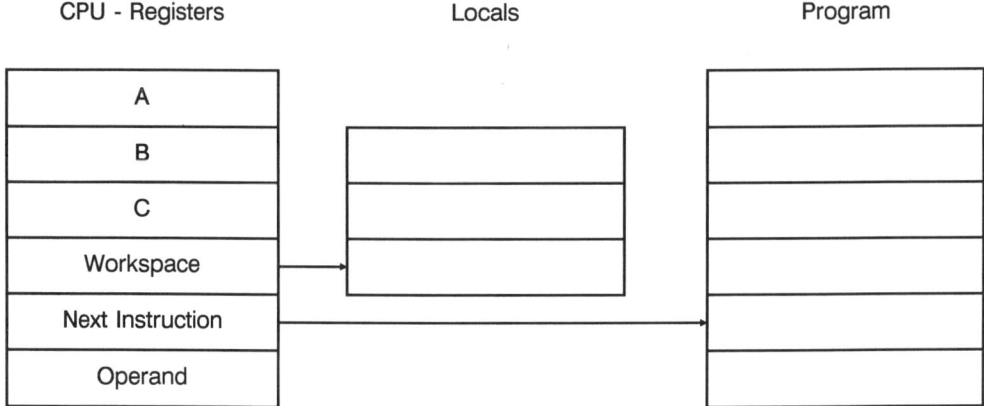

Figure 2: CPU - Registers

Finally, the CPU also contains the two registers Front and Back, both controlling a linked list of processes waiting to be executed. As their names indicate, these registers point to the start and end of the waiting list, which is internally linked by pointers, also. More details about this feature will follow.

INSTRUCTION FORMAT

One major design goal of the Transputer hardware and instruction set was compatibility and optimization for use with high level programming languages. Furthermore, the Transputer customized language OCCAM largely eliminates any need for Assembler programming. The ability to also program the Transputer in common languages such as C, Pascal or FORTRAN allows use of a large selection of existing software, an advantage not to be underestimated. Special features of the Transputers - parallel processing or communication via links - are made accessible to these languages through appropriate library extensions.

The question of whether the Transputer is a RISC (= reduced instruction set) or a CISC (= complex instruction set) processor may never be completely resolved, the fact is it combines the best of both worlds. A new, innovative coding procedure allows the instruction codes of the Transputer to be of different lengths. Indeed, instructions most frequently used when programming in high level languages like OCCAM or C use the least space in memory. Tests have shown for example, that Transputers spend 80 percent of their time executing a kernel of only 31 instructions (5).

As shown below this instruction coding technique has an intended side effect: instructions are independent from the processor's word length (16/32 bit) (1).

The basic Transputer instruction always consists of only one byte which is divided into two nibbles, four bits each. The four most significant bits form the instruction code, the four least significant bits a date (see figure 3).

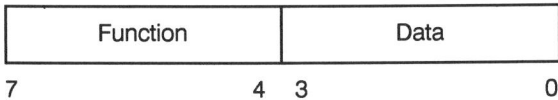

Figure 3: Instruction format

Basically there are three groups of instructions: direct, indirect and prefix instructions. In direct instructions the most significant four bits of a byte contain the instruction itself, whereas the least significant four bits are loaded into the operand register as data, where they can be accessed by the instruction. Therefore these four bits provide the coding of 16 different direct instructions, of which 13 of these are the most frequently used instructions in high level languages. A range of 0 to 15 for data might seem rather small at first glance but is sufficient for an astonishing variety of applications. For example very often small numbers have to be loaded or stored for loops and just as often these four bits are sufficient to address variables relative to the workspace pointer or to the address contained in register A. Table 1 lists function codes, mnemonics, number of cycles required and the names of these 16 instructions.

Two of these 16 direct instructions are special, they are called prefix instructions and provide the ability to generate higher numbers. They always write 4 bits of data into the operand register but carry out no further operations. The prefix instruction first writes the 4 bits of data into the lowest four bits of the operand register and then shifts the contents of the operand register four bits to the left. Numbers of any size (to a maximum of 16/32 bits) can thus be generated in the operand register by combining several prefix instructions. The negative prefix instruction works like the prefix instruction but complements the operand register before the left shift. This provides an easy way to produce negative numbers. Table 2 shows some simple examples for handling prefix instructions.

The use of prefix instructions brings certain advantages to the coding. For example arguments can be generated independent of the word width of the CPU (of course 32 bit words can not be generated in a 16 bit CPU). Also the compilation of high level languages is simplified since the generation of arguments remains identical for all operand instructions.

Table 1: Transputer direct function codes

Code	Mnemonic	Cycles	Description
0x	j	3	jump
1x	ldlp	1	load local pointer
2x	pfix	1	prefix
3x	ldnl	2	load non local
4x	ldc	1	load constant
5x	ldnlp	1	load non local pointer
6x	nfix	1	negative prefix
7x	ldl	2	load local
8x	adc	1	add constant
9x	call	7	call
Ax	cj	2	conditional jump (not taken)
		4	conditional jump (taken)
Bx	ajw	1	adjust workspace
Cx	eqc	2	equals constant
Dx	stl	1	store local
Ex	stnl	2	store non local
Fx	opr	-	operate

Table 2: prefix coding

Mnemonic	Function code	Memory code
ldc #35		(is coded as)
pfix #3	#2	#23
ldc #5	#4	#45
ldc #789		(is coded as)
pfix #7	#2	#27
pfix #8	#2	#28
ldc #9	#4	#49
ldc -31		(is coded as)
nfix #1	#6	#61
ldc #1	#4	#41

The operate instruction, the last of the 16 direct instructions, interprets the operand register's contents as an instruction code. Since, in the case, the 4 data bits of the instruction byte are interpreted as function code, a further 16 instructions are available with just one single byte sequence. If needed, the contents of registers A, B and C of the CPU stack may be used as arguments for these instructions. Figure 4 and table 3 together give two short examples for the use of the operate instruction.

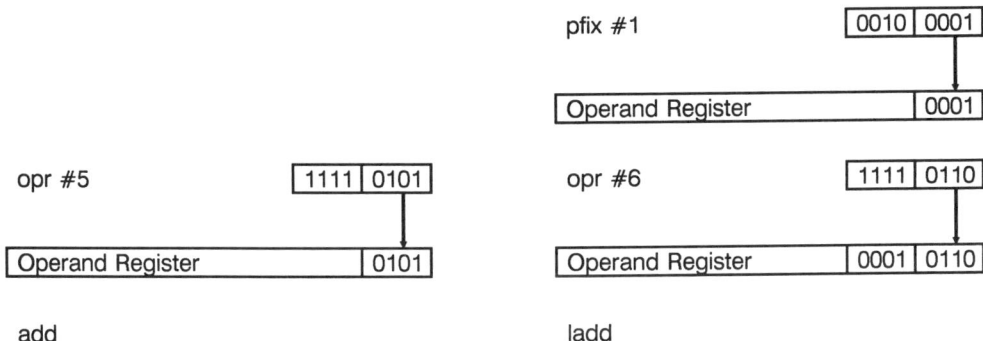

Figure 4: Operand register contents for the two examples in table 3

Table 3: Examples for operate coding

Mnemonic	Function code	Memory code
add opr #5	#5 #F	(is coded as) #F5
ladd pfix #1 opr #6	#16 #2 #F	(is coded as) #21 #F6

The 29 most frequently used instructions can therefore be coded with only one byte (32 minus prefix, negative prefix and operate). According to INMOS, results from a large number of benchmarks have shown that one byte coding is sufficient for an average of 70% of the instructions used within a program, with most instructions being executed in only one cycle. Since memory is accessed by words (16/32 bits), a further advantage of such short instruction sequences is, that in 32 bit processors four instructions are available at one time. Together with a double buffered instruction queue this instruction prefetch delivers the same advantages as a cost intensive silicon instruction cache. It also decouples instruction execution time from memory speed to a large extent.

Well above the RISC level, Transputers have more than 160 instructions (depending on type), generated by a prefix and an operand instruction, some of which trigger a longer sequence of microcoded functions providing support for complex high level routines. In particular, instructions for the handling of parallel processes, for communications, graphics and complex mathematical operations are noteworthy. They make compiled programs for Transputers less than half size of programs for typical RISC processors. Table 4 shows a sample of instructions and their realization as combinations of prefix and operate instructions.

Table 4: Examples for Transputer indirect function codes

Operation	Memory code	Mnemonic code	Cycles	Description
46	24F6	and	1	A = B and A
4B	24FB	or	1	A = B or A
41	24F1	shl	n+2	B = B << A
05	F5	add	1	A = B + A

PARALLEL PROCESSES

The most significant disadvantage of having a large set of CPU registers is the time required for context switching, i.e. the time required to save the registers in the event of an interrupt and to reload them afterwards to allow continuation of the original program without loss of information. To reduce this time overhead for context switching as much as possible, the Transputer's hardware is designed as a multitasking machine. In conventional processors periodical switching between tasks or processes is realized in software, giving multitasking operating systems a slow response, whereas on a Transputer this is realized as a microcoded, fast hardware scheduler.

Concurrent processes running on only one Transputer are executed via context switching, also. A process can have one of the following two conditions:

        active   process is being executed
                          process is waiting in the process list to be executed

        inactive   process is waiting for I/O
                          process is waiting until a specified time

In addition to these two conditions there are two priority levels. Priority-0 processes have higher priority, this means they are always executed preferentially. As soon as a high priority process is to be started, a priority-1 process is interrupted and is not started again as long as any other high priority process is still active. The priority of a process is coded in bit 0 of the workspace address, which is always adjusted to a word bound (16/32 bits).

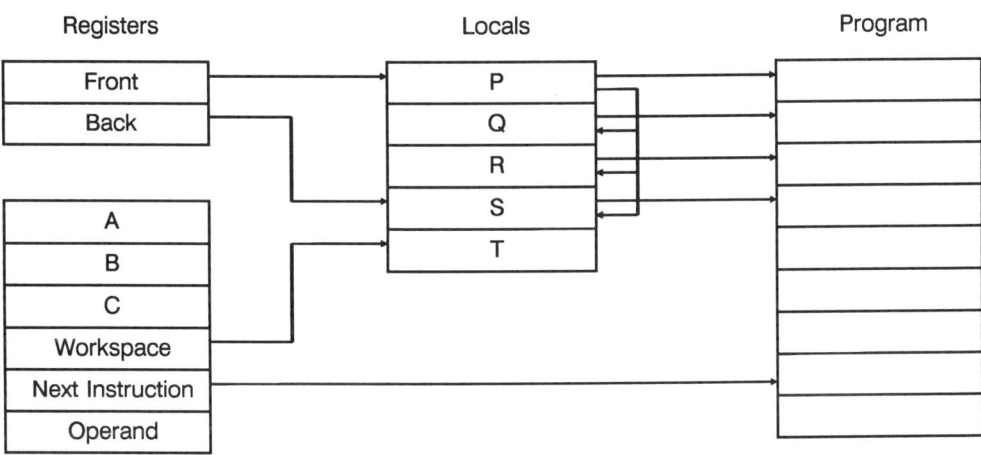

Figure 5: Linked process list

Inactive processes do not require any CPU time and active processes (those which receive CPU time in the the time slice process) are controlled in a linked list. The Transputer accesses this process list via registers Front and Back as mentioned above. As shown in figure 5 the front pointer points to the beginning and the back pointer points to the end of the process list and they themselves are linked together by pointers. Processes of different priorities are managed by different Front and Back pointers as well as different process lists.

A process will stay active (maintained in the process list) until it waits for I/O or a time slice. If an executing process is interrupted, the actual value of the instruction pointer is saved at a predefined address of the corresponding workspace and the next process in list continues its execution. As registers A, B and C aren't saved in a process switch, such a switch is only possible at instructions that guarantee no detrimental effects as a result to changes of registers A, B or C. If an inactive process is activated, it is linked to the end of the process list and the Back pointer is modified accordingly. As a result of hardware process scheduling and the low number of registers to save, context switches are extraordinarily fast, requiring less than one microsecond. In comparison, a 20 MHz 80386 spends about 17 microseconds and even a 20 MHz SPARC still needs 15 microseconds. A T800 running at 30 MHz requires only 0.63 microseconds for a task switch and a 20 MHz T800 needs less than one microsecond (5).

CHANNEL COMMUNICATIONS

Although context switches are fast, a single Transputer cannot really execute processes in parallel, i.e. simultaneously, but has to execute them time sliced, sequentially as indicated above. Communications via links is, however, independent from the CPU and thus can be performed in parallel to program execution.

The best approach to execute parallel processes simultaneously is to employ several Transputers with the individual processes assigned to different processors. The number of processes is however not limited to the number of Transputers, as different processes can be executed on one Transputer "quasi parallel". An essential prerequisite for such a multiprocessor network is efficient communications between the processors. Conventional processors running in such multiprocessor networks generally lack such efficient communications capabilities. Typically, data transfers have to use a bus, which becomes the limiting factor in determining the overall system performance as the number of processors increases.

With exeption to some special implementations for mainframes (e.g. Suprenum), the Transputer's concept of channel communications offers one of today's most efficient solutions for communications in a multiprocessor network. As indicated in its name, communications between parallel processes are executed via channels in Transputers. These channels are point-to-point synchronized and unbuffered links.

There are internal channels providing communication among processes within a single Transputer and external channels for communications between processes running on different Transputers. The internal channels are represented by a single word in memory, whereas external channels are implemented as serial transmission highspeed links, fixed to predefined memory addresses. The present Transputer generation, with a maximum of four links, imposes certain limitations to building complex Transputer networks. As a result not all Transputers may be able to directly communicate with each other, but may have to route through one or more other Transputers. This is usually a serious hindrance only in large network systems, since the communication is largely independent of the processor's other work.

Internal Channel Communication

The memory address assigned to an internal channel contains either the special identifier 'empty' or the workspace address of a process. Before any communication the internal channel must be on its initial 'empty' state. A process wanting to communicate is then interrupted. Its workspace address is put into the internal channel and finally the process is switched inactive, i.e. it is removed from the process list. When the second partner for this communication is ready and wants to access the activated channel, data in memory is transferred with a simple block move instruction. Having finished the transfer, the inactive process is picked up into the process list again.

Figures 6 to 9 illustrate the flow of an internal channel communication between the two processes P and Q.

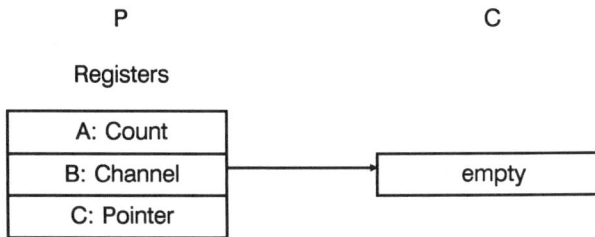

Figure 6: Output to empty channel

Initialization of an internal channel presets channel word C with 'empty' (=0x80000000), signaling a free channel. At this time the stack has to contain a pointer to the data in register C, a pointer to the channel word in register B and the number of bytes to transfer in register A (see figure 6). After the output instruction channel C holds the workspace address of P, where the number of bytes and the data pointer are located at predefined addresses in the workspace of P (see figure 7). Finally process P is interrupted and switched inactive to continue execution of the next process in the process list.

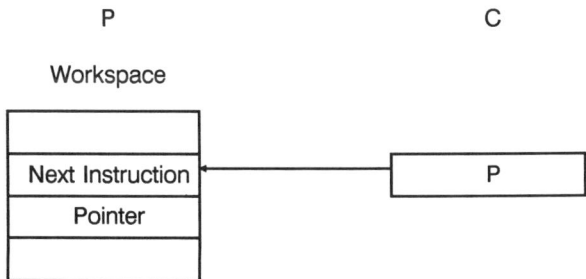

Figure 7: Channel and workspace after output instruction

Channel C and process P will remain in that condition until another process Q wants to execute an I/O instruction on this channel (see figure 8). The data is copied, P is returned to the process list and C is reinitialized with 'empty' (see figure 9).

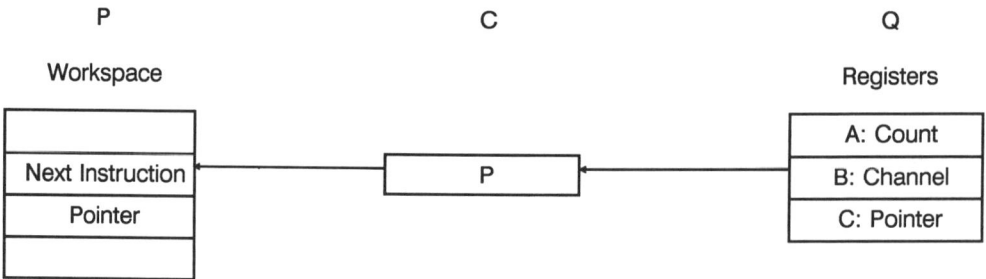

Figure 8: Data transfer between processes P and Q

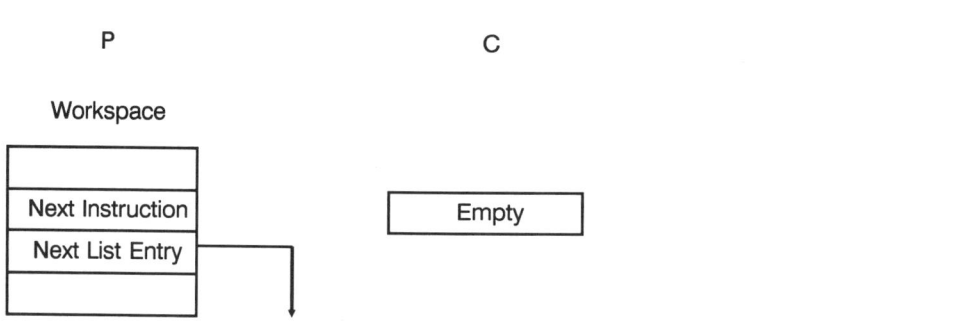

Figure 9: Channel and workspace after communication

It is prohibited to have more than one process obtain read or write access to the same channel. If for example two processes tried to read from a channel simultaneously the process which was ready last would expect its partner to be a writing process and data would be copied wrong. Clearly, these communication rules are optionally suited for the synchronization of processes.

External Channel Communication

If a process P wants to communicate via an external channel - one of the Transputer links - with a process Q on another Transputer, the CPU hands the execution of this task over to a completely autonomous link interface. This supervises the process interruption, the switch to inactive status and the return to the process list once the data transfer is complete. Each link interface has its own DMA mechanism, capable of transferring data between link and memory without requesting the CPU. Each link interface uses three registers to take the following three values:

    a pointer to the process workspace

a pointer to the data
the number of bytes to transfer

Figures 10 to 12 show how two processes P and Q on two Transputers T1 and T2 communicate, each via links 0. P sends the data which Q receives.

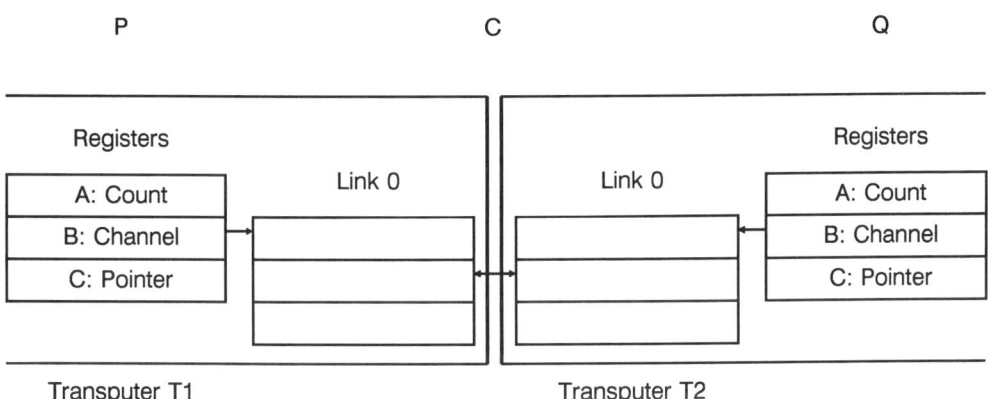

Figure 10: Communication between transputers

Figure 10 shows the contents of the CPU before the output instruction. As soon as P executes that instruction, the three registers of link interface 0 in T1 are initialized as shown in figure 11, and P is interrupted. The procedure with Q after the input instruction is analogous (see also figure 11). Only if both processes are ready - independent of the occurrence - data transfer is executed from P to Q. Finally both processes are returned to the process list from their link interfaces (see figure 12).

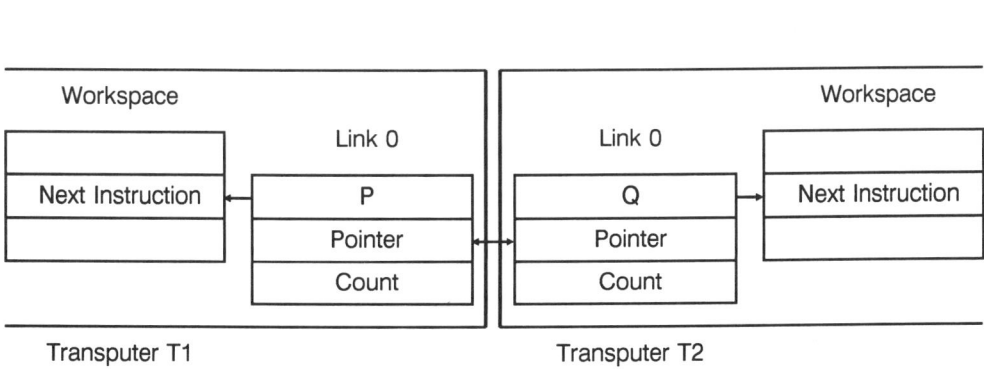

Figure 11: Workspaces and link interfaces after output instructions

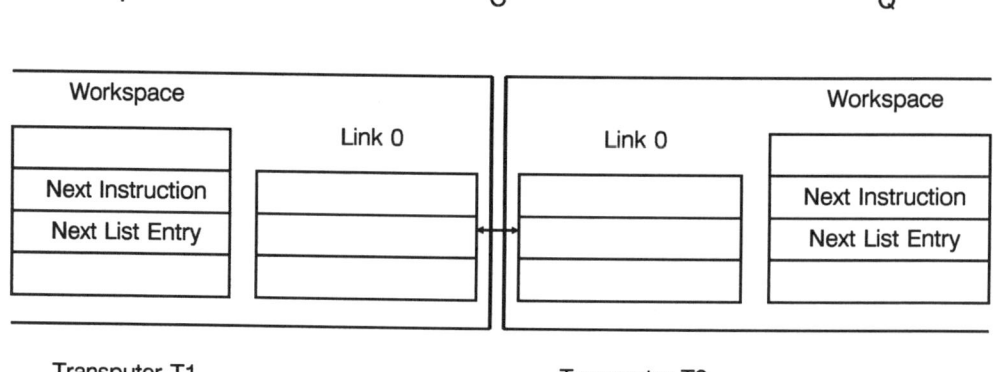

Figure 12: Transputers after communication

Transputerlink

A link connection between two Transputers consists of a two wire connection between the Transputers' link interfaces. Serial signals are transmitted along each of the two uni directional leads, with a hardware selectable transfer rate of 5, 10 or 20M bit/s. Equally simple is the transfer protocol, illustrated in figure 13.

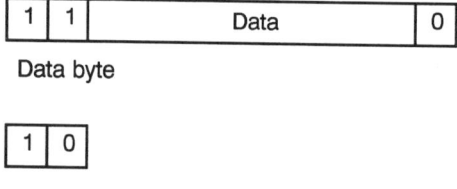

Figure 13: Link data and acknowledge format

The transfer begins with a start bit followed by a bit with logical value 1. The eight data bits and a stop bit then follow. Acknowledgement on the second lead is a start bit followed by a stop bit, signalizing the process is ready to receive and that there is buffer available for a further byte. This protocol allows the acknowledgement to be sent before the data byte is completely transferred, thus providing a near continuous transfer without the need to wait for acknowledgement after each byte. This overlapping acknowledgement results in a transfer rate of 1.8M bytes/s. The acknowledgement on some Transputers (e.g. T414) is non overlapping, yielding an average transfer rate of 0.8M bytes/s.

The data format of the transfer resembles that of the asynchronous RS-232 protocol. Bytewise transfer and acknowledgement of data makes transfer via links independent from the Transputer's word width, permitting all Transputers to directly communicate with each other via links.

PERFORMANCE COMPARISONS

In Technical note 27 with the appropriate title "Lies, Damned Lies and Benchmarks", INMOS has published some interesting comparisons of Transputers and other processors. The authors place particular value on the uniformity of test conditions when performing processor benchmark tests, without which the result can be a misrepresentation. This is particularily relevant for manual or compiler optimized code segments, which may over or underestimate processor specific features. The following two widely used benchmark programs were used for the present purposes.

Whetstone Benchmark

The Whetstone benchmark program measures a processor's power in scientific applications. Besides pure floating point operations it also contains integer arithmetic, array indexing, function calls, conditional branches and basic operations like, for example, commutations.

Table 5: Whetstone benchmark comparison for single and double precision

System	Thousands of single precision Whetstones per second
IMS T800-30	6400
IMS T800-20	4000
WE 32200/32206-24	2800
INTEL 80386 + 80387	1860
VAX 11/780	1083
SUN-3	860
NS 32332/32081	728
IMS T414-20	667
INTEL 80286 + 80287	300
INTEL 8086 + 8087	178
MC 68000	13

System	Thousands of double precision Whetstones per second
IMS T800-30	3800
IMS T800-20	2500
INTEL 80386 + 80387	1730
SUN-3	790
VAX 11/780	715
IMS T414-20	163
INTEL 8086 + 8087	152

Dhrystone Benchmarks

Although the Dhrystone benchmark is as widespread as the Whetstone benchmark it is less well suited to measure a processor's speed because it uses only integer operations, nearly no looping constructs but therefore measures string operations which rely heavily on the hardware and the software used. Despite this the results have been presented below in Table 6.

Table 6: Dhrystone benchmark comparison

System	Dhrystones per second
IBM 3090/200	31250
IMS T414-20	8547
VAX 8600	6423
INTEL 80386-16	4300
MC 68020-17	3977
INTEL 80286-10	1976
VAX 11/780	1650
MC 68000-8	1136

It should be noted, that many Dhrystone figures, especially those quoted by manufacturers, are often invalid. Either they refer to the incorrect version 1.0 (and if no version is given, this is usually the case) or else they use optimising compilers, which are forbidden for this benchmark (frequently both). The figures above are believed to be free of such contamination (3).

CONCLUSIONS

The INMOS Transputer family forms a standardized line of modules for parallel, state-of-the-art computer systems. The high performance of the Transputer, a simple modular design, modest requirements for support circuitry even for special applications, quite simple ways to program even very complex multiprocessor networks and simple communication between processes have led to a fast rising popularity of Transputers. Since becoming a subsidiary of SGS-Thompson Microelectronics, the Transputer development has been given a top priority at INMOS, and thus the future of Transputer is well assured. Today there already exists more than thousand of Transputer applications, e.g. in robotic control systems, in picture and signal processing, in artificial intelligence and recently even in neuronal networks.

References:

1 The Transputer Databook,
   INMOS Ltd., 1989, document number : 72 TRN 203 01
2 The transputer instruction set - a compiler writers' guide,
   INMOS Ltd., 1988
3 Roger Shepherd, Peter Thompson,
   Lies, Damned Lies and Benchmarks,
   Technical note 27,
   INMOS Ltd., 1988
4 Gerd Häußler, Peter Guthseel,
   Transputer,
   Franzis Verlag, Munich 1990
5 Peter Eckelmann,
   Transputer, was sonst?
   Begleittext zum 7. Entwicklerforum "Transputer",
   Markt & Technik Verlag, Munich 1989
6 Gerd Häußler,
   Die Assemblersprache des Transputers
   MC Mikrocomputer-Zeitschrift 6/88 and 7/88
7 Uwe Hildebrand,
   Die mc-Transputerkarte
   MC Mikrocomputer-Zeitschrift 5/88 and 6/88

Transputers in Technical Applications

Dr. Siegfried Streitz
PARACOM GmbH, Jülicher Str. 338, D-5100 Aachen

INTRODUCTION

Parallel processing is rapidly becoming the mainstream of new data processing developments, taking over from the sequential von Neumann type of computer. Transputer based systems demonstrate the practical feasibility of supercomputer performance integrated with worldwide standards. This is done by means of bus interfaces supporting parallel processing within existing systems, workstations like VAX, PC, PS/2, SUN, VME, MAC, etc. and by making available standard programming languages such as FORTRAN, C, etc.

As an example of a large supercomputer system used in industry and purchased in a commercial environment, a 400-processor transputer system will be described.

400 TRANSPUTER SUPERCLUSTER FOR SHELL

In the Mathematics and Systems Engeneering department at 'Koninklijke Shell Laboratorium Amsterdam' (KSLA), a parallel computing research project was started in 1985. Parallel computing at KSLA and the Transputer technology in particular, have resulted in a completely new approach to fluid flow simulation called Cellular Automata. For this research, there is a need at KSLA for a massive parallel system.

PARSYTEC from Aachen, Germany, has been chosen to supply a 400 Transputer system. PARSYTEC was one of the first companies worldwide to accept the Transputer concept and put it into parallel computers. It now is the largest supplier of industrial Transputer systems worldwide.

REQUIREMENTS

- Supercomputer performance at the best price

- Scalability to adapt performance to problem size

- Highest Reliability and Fault Tolerance

- Decdicated total computing power for industrial applications

U. Harms (Ed.)
Supercomputer and Chemistry
© Springer-Verlag Berlin Heidelberg 1990

Supercomputer performance is required when simulating the basic mechanisms underlying fluid flow phenomena. An approach, called Cellular Automata, is more direct than the traditional approach of solving partial differential equations. One major advantage of Cellular Automata is the high speed and low cost of the computations on parallel computers.

Scalability means the capability to add new processing power when the problem size increases or to speed up an application by adding further processors. Research at KSLA has proven that all KSLA applications including Cellular Automata can efficiently run on a computer network configured in a mesh topology. Initially, a 20x20 mesh is required. Further expansion of the system is planned and should in no way be limited by physical constraints.

Reliability is naturally expected in every system, but the requirements to achieve it grow with the size of the system. To guarantee highest reliability for massively parallel systems, especially for time consuming applications, each component has to be designed very carefully. Fault tolerance is essential and is inherent into parallel systems.

Dedicated total computing power becomes more and more important for time consuming applications. For example the actual portion of CPU time on traditional supercomputers depends on the load on the system. One can never predict whether an application requiring 4 hours of CPU time will run in 8 hours, one day, a weekend or more.

PARALLEL ARCHITECTURE WITH TRANSPUTERS

Transputer systems are based on an ingenious, but nevertheless simple concept:
- processor
- coprocessor
- memory
- communications channels

on **one** chip.

The construction of multiprocessor systems is facilitated by the concentration of all important resources in one element. Adding more nodes means adding channels and therefore communications bandwidth, so the overall balance of computing power and communication bandwidth is always maintained. The interprocessor scheme breaks the bottleneck imposed by convential bus-based or other memory-coupled designs.

Transputer systems offer a promising approach to parallelism for research, teaching and industry. Performance can be increased in small graduations to meet computational demands. Program development only needs one or two transputers, so implementation work can be performed occupying a minimum of hardware resources. Production runs are realized with multiprocessor configurations depending on the problem size.

PARSYTEC has a company mission to extend the transputer features to the design of a complete parallel system. The high-end of the product line is represented by the sophisticated SuperCluster series.

SUPERCLUSTER ARCHITECTURE

The SuperCluster series has a hierarchical, cluster-based architecture. The basic unit of the system is a computing cluster which consists of 16 processors. Two types of clusters can be distinguished: the Variable Topology Cluster and the Regular Topology Cluster. The first type of cluster can be configured to any topology required by means of a Network Configuration Unit (NCU). The second type forms a 4 X 4 matrix, which is the basic building block for applications, which require a fixed and regular structure of processor topology.

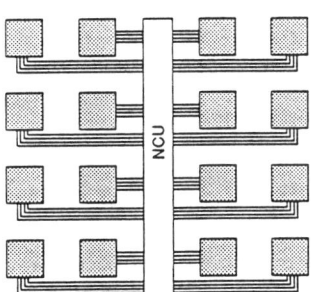
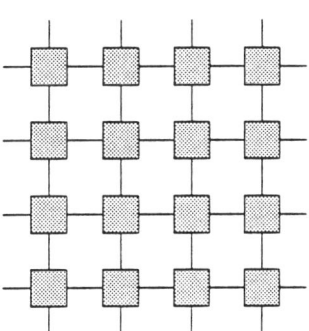

Variable Topology Cluster   Regular Topology Cluster

The next level is a set of clusters, which can again be interpreted as a cluster, and which, on the whole, enables a more powerful realisation of processor resources and the execution of complex application functions. For the reconfigurable type this means that two computing clusters (each including one NCU) are interconnected again by another NCU. Thus, for example, four computing clusters and two NCUs form the basic Supercluster unit, the Model 64. The clusters of the Regular SuperCluster are interconnected by cables, which link the "sides" of the 4x4 matrices.

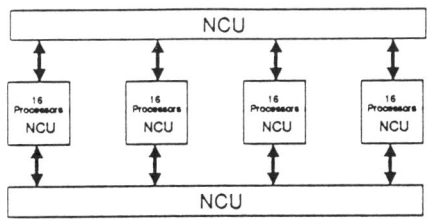
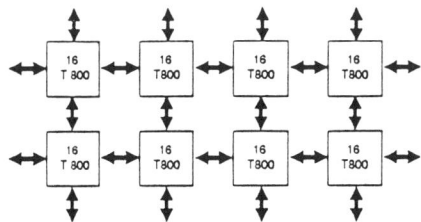

Variable Topology SuperCluster (Model 64)   Regular Topology SuperCluster (Model 128)

Multiple units again form a larger systems. The Model 256, for example, has four basic units which are linked through their NCUs. Twenty-five Regular Topology Clusters form the 400 processor Shell system, configured in a 20x20 mesh. A 32x32 mesh topology, configured out of 1024 Transputers is simple a next step. The concept described here shows the modular expandability and uniform behaviour throughout the system. This is in sharp contrast to more conventional super computer architectures which often cannot be upgraded to meet increased demands.

A basic unit may also include a System Services Cluster for parallel mass storage systems, user interface modules and application-specific interface modules. All of these are based on the same transputer-aided communications technique as the application processors.

The SuperClusters are operated by frontend workstations, which are interconnected by using the Transputer links. Interfaces are available for PC, PS/2, SUN, Macintosh II, VAX and VME. The maximum link speed amounts to 20 MBit/sec at distances of up to 10 m. A slightly reduced transfer rate is achieved at distances of 10 to 30 m.

The basic processing module of a SuperCluster is the MTM-EDC. Four transputers are assembled on each MTM-EDC. Each transputer has a local storage capacity of up to 4 MBytes and several EDC units (Error Detection and Correction units) which at intervals of 4 seconds automatically check the complete storage for bit errors and correct them when necessary. Furthermore these units check and correct, if necessary, every word when it is read from memory. Each cluster is represented by four MTM-EDC modules.

The Shell system has a fixed configuration to form a 20x20 torus mesh. Therefore the MTM-EDC's are configured as a 2 x 2 processor matrix. Each processor is supplied with 2 Mbytes of DRAM. Four modules with a total of 16 nodes give a 4 x 4 matrix, hardwired by a backplane element (BLE).

The interconnections between clusters are given by 32 way (corresponding to 4 links) flat cable connectors, which connect the 4 "sides" of the matrix. Furthermore the backplane has to reflect the different kind of link signals. Internal links and reset lines are TTL level, the 16 external links and reset lines are RS-422 buffered. Cabinet interconnection using RS-422 lines is guarenteed by conversion piggybacks.

CONCLUSIONS

The 400 processor Shell system produces 4000 MIPS or 600 scalar-MFLOPS, which relate to about 3 vector-GFLOPS on conventional supercomputers. Even more important is the potential for unlimited system build-up. Already in December 1989 the system will be installed with 100 processors, which will be upgraded in 1990 to 400 processors, still leaving the possibility to expand it to 1000 or more processors.

Plugging additional Transputer nodes into a system leads to a simultaneous increase in computing power and communication bandwidth. Each node adds four point-to point communication channels operating independently of other parts of the system. With this balanced approach almost unlimited performance levels can be achieved even when using hundreds of processors.

Reliability and fault tolerance are assured by the Transputer as it is a complete computer on one chip. In computer networks the Mean Time Between Failure (MTBF) decreases exponentially with the size of the network and memory. Error Detection is essential, but for a 400 processor / 800 M-bytes system not enough. PARSYTEC transputer modules are therefore equipped with Error Detection and Correction on all local memory segments, resulting in a theoretical MTBF which is far beyond the lifetime of any computer system with G-bytes of memory.

Another source of faults, communication link errors, have been minimized by utilizing a differential driven communication scheme (RS-422) between all Transputers located on different modules. This concept makes the system almost insensitive to electromagnetic radiation. PARSYTEC systems are air-cooled and therefore do not require extensive (and expensive) cooling systems such as those based on liquid nitrogen.

Quantum Chemical Calculations on MIMD Type Computers

U. Wedig, A. Burkhardt, and H. G. v. Schnering
Max-Planck-Institut für Festkörperforschung, D-7000 Stuttgart 80

Abstract: MIMD parallel computers can be a promising alternative in those cases where application software exploits only a small portion of the maximum performance of vector computers. We compare MIMD systems with other parallel architectures and stress transputer networks pointing out the benefits of this concept. Two different software environments are discussed, MultiTool, a modified transputer development system (TDS) and the distributed operating system Helios. Both are used to implement direct SCF programs on a transputer network. Parallel processing is defined according to the farming concept, both on the program and on the subroutine level. In transputer networks with up to 17 nodes, a very good load balance could be achieved, demonstrating the good suitability of the MIMD-concept for SCF calculations.

INTRODUCTION

The calculation of the structures and properties of molecules and solids by ab initio methods requires an enormous amount of computer performance. The effort increases with up to the fifth power of the number of basis functions, which depends on the accuracy desired and on the size of the problem treated. Towards higher accuracy and larger systems three ways are possible. We may develop new methods, improve our algorithms and programs or look for more powerful computers. Actually we have to investigate all three aspects if we want to apply successfully the best hard- and software to perform quantum chemical calculations.

This article deals with the third aspect. In the next section we summarize the principles of computers being used in the last decades to lead over to some possible parallel computer architectures. We discuss various characteristics of parallel systems like the capabilities of the processing units, the way how they are connected and the inter-processor communication. Parallelism may occur on the instruction level or by distributing complex tasks in a processor network. In addition we explain why we use transputers for our investigations.

A supercomputer is nothing but some dirty silicon combined with some plastics and some metal. It is only valuable in connection with good software. We need a reliable operating system and tools to develop and run application programs easily. For transputer systems two types of software environment are available. In the third section we outline concepts how they can be used to develop a distributed program.

Some procedures used in quantum chemistry have been implemented on vector computers very successfully. Examples are the transformation of the two electron integrals from an AO to a MO basis or the direct CI method [1]. However, this is not true for other algorithms. The Hartree-Fock method with no restrictions concerning the basis set is one of the most important of them. (cf. [1,2,3,4]). MIMD parallel computer are more suited for this typ of calculations. The decomposition of an SCF program into tasks, which may be processed concurrently, is described in the fourth section.

COMPUTER ARCHITECTURES

Single Processor Architectures

Until now, most computers are scalar machines, using only one central processing unit (CPU). Their method of operation is shown schematically in Figure 1a. The operands have to be loaded into registers of the CPU, and are processed in a functional unit. The result is stored via a CPU register in the memory of the computer. The operation takes several processor cycles, depending on the number of steps in the functional unit.

The hardware of the CPU can be utilized more efficiently if more than one register set is used, and several groups of operands are processed simultaneously in various parts of the functional unit. This kind of overlapped processing is called *pipelining*. Pipelining is the basic idea of vector computers. Equivalent operands are grouped in vectors and stored in vector registers of the CPU. Figure 1b shows the simplest form of a vector processor. In a first processor cycle the first element of vector A (A1) and the first element of B (B1) are processed in the first part of the functional unit (F1). In the second cycle A1 and B1 are processed in F2, A2 and B2 in F1, and so on. After a start-up time, required to fill up the pipeline, a result is produced every processor cycle and stored in vector register C.

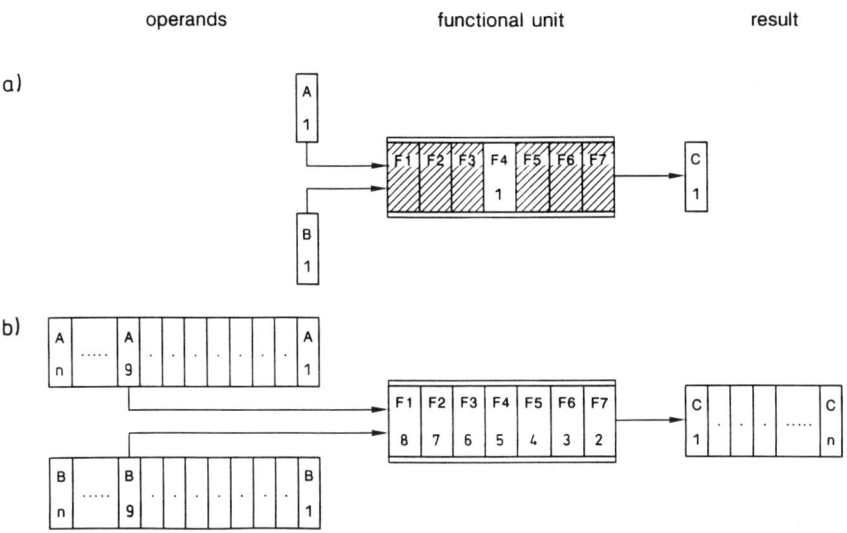

Fig. 1: Functional scheme of scalar (a) and vector (b) processors.

The vector architecture, although offering a new dimension of computing power, has some disadvantages. In order to achieve maximum performance for a simple addition of two vectors, the memory has to be accessed three times (two read and one write) during one processor cycle. This requires a very sophisticated management of very fast memory units and very fast data paths. The fact that high end technology has to be used for all the components is one reason for the high price of vector supercomputers. Beyond this the hardware is only worth the money if algorithms with appropriate data structures run on it. However, there are many important computational methods which, until now, couldn't be vectorized satisfactorily.

Parallel Computer Architectures

In the last decade, the peak performance of a CRAY processor was increased from 150 MFlops (CRAY-1) to 333 MFlops (CRAY Y-MP) or 490 MFlops (CRAY-2). At the same time the computing power of microprocessors grew by nearly two orders of magnitude. At the high end, the performance seems to converge. Substantially more computer power will only be available if several processors run concurrently. Future supercomputers will be parallel computers, but, what will they look like? The summary of the results of a workshop on parallel algorithms and architectures [5] is the only answer that can be given today.

We can call it 'parallel processing', but we don't know how to do it. We do need experiments, however, to suggest possible new directions.

The uncertainty concerning the concepts in parallel computing is reflected by the large variety of types of parallel computers available on the market. Among other things, they differ in the complexity of the processing units (PU), in the way the PUs are connected and in the granularity of the tasks running in parallel.

Processing units: Their complexity ranges from simple processing elements (e.g., connection machine, DAP) over microprocessors (e.g., transputer networks, Intel iPSC, Suprenum) to super vector processors (e.g., CRAY, IBM). Lower performance of the single unit can be compensated by a larger number of them. The highest performance/price ratio can be expected from systems built of mass products like microprocessors. However, as mentioned later, it is more difficult to write distributed programs for this class of computers.

Connections: Different ways of connecting processing units are shown in Figure 2. Closely related to this is the method how the processors exchange information. In global memory systems (Fig. 2a) this is done by access to the same memory location. Limited by the bandwidth of the memory, they are restricted to a small number of PUs. Like single processor vector computers (v.s.), they need very fast and therefore expensive memory and data path components.

In distributed memory systems (Fig. 2b,c), every processing unit is connected to its own local memory. The nodes (PU + memory) communicate by *message passing*. Messages may be sent from one node to another via a common bus (Fig. 2b). The maximum transfer rate of the bus limits the number of nodes attachable.

Much larger parallel systems can be built up by connecting the nodes to a multidimensional network (Fig. 2c). Only the communicating and the intermediate nodes are involved in passing a message. Most of the other nodes are not affected by the data transfer.

Granularity of the tasks: The tasks being processed on a single node may vary from simple instructions to complex programs. Generally, in the SIMD concept (Single Instruction, Multiple Data), the same instruction is performed in a certain number of PU's, each using another set of operands. Sometimes, vector computers are also considered to be SIMD machines, although the identical operations are not performed in the same cycle. The elements of the vectors, however, are processed independently. MIMD (Multiple Instruction, Multiple Data) means, that larger

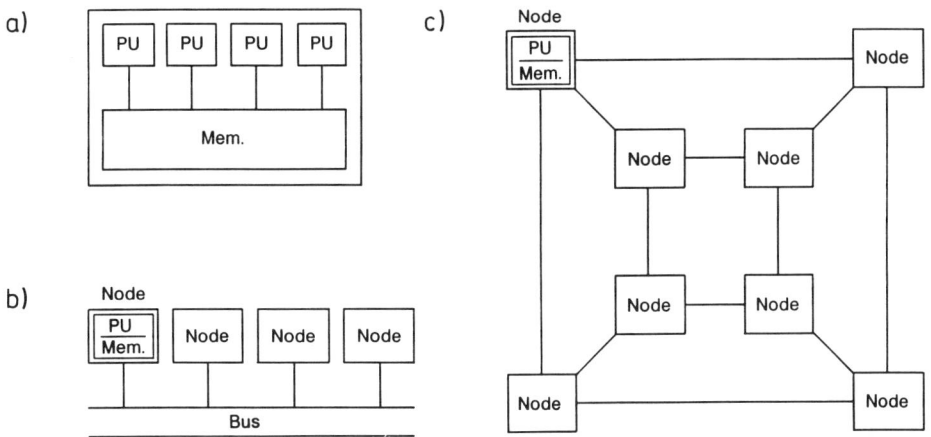

Fig. 2: Connection of processing units (PU) in parallel systems.
   a) global memory
   b) local memory, bus system
   c) local memory, multidimensional network

tasks are distributed among the PUs. If the parallelly running tasks perform the same operations, the term SCMD (Single Code, Multiple Data) is used sometimes.

A given parallel hardware does not necessarily imply the use of one single concept. For example, on parallel vector computers both MIMD and SIMD structures are applicable. In the latter case, vector operations are subdivided and the parts are processed on different PUs (e.g., CRAYs autotasking).

When using the SIMD concept, we have to apply the same rules as for the vectorization. As a consequence SIMD is not suited for algorithms which cannot be vectorized well. The algorithms have to be programmed in a MIMD way. For that purpose, the best hardware platforms are concurrent microprocessors with a sufficient amount of local memory. They are able to process complex programs and offer the highest performance/price ratio.

Transputer Networks

The transputer is a whole computer on one chip. The T800 model contains a 32-bit RISC processor, a floating point unit, local memory which may be expanded externally and four I/O channels. It was designed to build up multidimensional networks of processors [6]. Simultaneously, INMOS developed the programming language OCCAM [7]. In addition to many elements common to other sequential languages, OCCAM contains statements to control concurrent processes, communicating through unbuffered, unidirectional channels. The transputer and OCCAM are optimized for the use of the MIMD concept on the basis of message passing. The benefits of the transputer and OCCAM are described elsewhere in this volume. Detailed informations can also be found in some articles of Hey [8,9]. Recalling the quotation mentioned above to memory, we surely cannot proclaim that transputer networks are the only answer to the question of the best parallel computer concept. However, to gain experience in parallel processing, they are a very efficient and a good value starting point.

PROGRAMMING TECHNIQUES FOR TRANSPUTER SYSTEMS

When passing over to a new computer, the majority of the users prefers the following method: Take the old FORTRAN program, compile it and start it. This is possible of course with scalar machines. It may be possible with vector computers, when using an autovectorizing compiler. However, to get a reasonable performance it is often necessary to reorganize the code (e.g. DO loops) or the data structures. In several cases even this proceeding does not lead to a program with a satisfactory degree of vectorization. The statements concerning vector computers hold also for SIMD systems.

If we want to use the MIMD concept, the only way to avoid programming effort is to run the whole program on a single node. Programming work has to be done in any case, if parts of a program should be distributed among the nodes. Autoparallelizing MIMD compilers are not available and cannot be expected in the near future. The code has to be partitioned into independent tasks. In figure 3 a general scheme is shown. An example application will be discussed in the next chapter.

Farming

A certain number of independent processes may be distributed in two ways. On one hand the location of a specific task may be predetermined. In this case it is necessary to estimate the required processing time for each task in advance to achieve a good load balance for the whole system. On the other hand the distribution can be organized dynamically by sending a task to the next processor that idles.

A random distribution of tasks using the same algorithm but different input data is called farming [9,10]. If the number of data sets exceeds by far the number of nodes, this is an easy way to

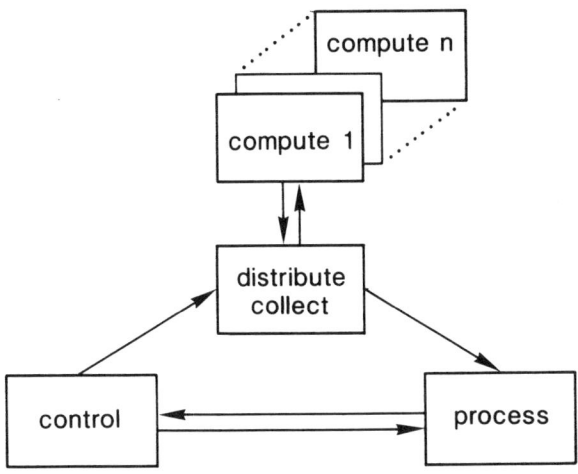

Fig. 3: A general scheme for the partitioning of a program into independent tasks.

achieve a good load balance. The advantage of farming compared to a predetermined distribution is drawn schematically in figure 4.

If the tasks are complete programs (farming on the program level), code written for scalar computers can be used with little modifications only. However, more costly nodes have to be used, as all of them must be equipped with the resources necessary to run the whole program (e.g. main memory, disks). Mechanisms are required to route data through the network to remote mass storage devices, if not all the nodes are connected to a local disk. As a consequence, the performance of the whole system may decrease. Farming on the program level leads to a high throughput of the system. However, the execution time of a single program will not be improved. Significantly shorter response times can be achieved if farming is used on the subroutine level. Programs calling complex subroutines within a loop are well suited for this concept, provided the maximum loop index is high compared to the number of processors available, and only a small amount of common data is shared between the subroutines. As these subroutines will run as independent tasks, common data and the parameter list have to be supplied by message passing. When using code from existing programs, the implementation of this communication is the essential part of the programming effort, needed to create a farm on the subroutine level.

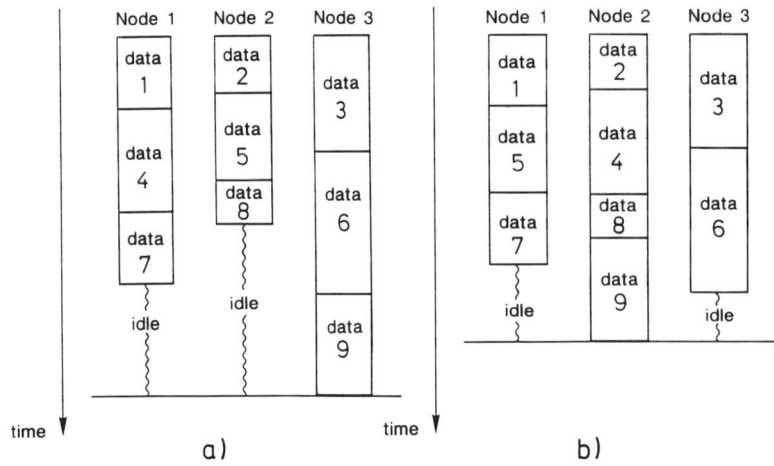

Fig. 4: The distribution of input data among concurrently running tasks.
a) predetermined distribution
b) farming concept

Programming Environments

Presently two completely different software environments are available to develop and run programs on transputer networks. The first group is based on the transputer development system TDS [11], supplied by INMOS. We used MultiTool, a modified version of TDS [12]. On the other hand there exist distributed operating systems with a UNIX like user interface. We consider Helios [13] as one example.

TDS, MultiTool: TDS is a software tool to edit, compile, configure and run programs. It's basic processes run only on one node. A server process, located on a host computer (PC, Sun workstation, ...) is communicating with TDS via a link adaptor. In general, TDS does not use the other nodes of the transputer network, therefore they are free for application processes. The distribution of tasks and the communication between them must be programmed by the user. The appropriate language is OCCAM, although FORTRAN and C compilers with extensions for the MIMD concept are also available. The integration of FORTRAN processes into an OCCAM harness is possible.

Helios: In contrast to TDS, Helios is a distributed operating system. A copy of the system

Example 1:   Task force defining a farm as outlined in figure 3.

```
component control
{
 code c_control;
 streams , , , , , , <| cntl1, >|cntl0;
}
component process
{
 code c_process;
 streams , , , , , , <| cntl0, >|cntl1;
}
control (| (|||[n]compute) | process)
```

nucleus is running on every node. This nucleus contains only the most important functions, the Kernel including the message passing algorithm for the basic interprocessor communications, the System Library, the Loader and the Processor Manager. Based on the client server model, other system services are rendered by server tasks which are located somewhere in the network, waiting for messages from other tasks requesting some operation or reply. The access to peripherials connected to network nodes or to an attached host computer is handled by such servers (e.g. [14]).

Helios offers an environment which is familiar to many users due to the UNIX like command set and a UNIX compatible library. The tasks (programs) can be written in standard programming languages. Extensions are not needed as the parallelism is not defined within the tasks themselves. To combine these tasks to one distributed program, a job control language (Component Distributed Language, CDL) is integrated into Helios. The tasks are components of a task force which will be interpreted by the Task Force Manager.

A task force, which applies the farming concept to the program outlined in figure 3, is defined by the statements given in example 1.

In the stream field of the component declaration, I/O streams are defined and connected to POSIX file descriptors. Not explicitly declared streams are allocated automatically. The last line defines the task force with the interleave constructor (|||). A load balancer process (lb) is inserted by Helios, sending the input data to the n *compute* processes and gathering the results. It corresponds to the *distribute & collect* process in figure 3. As the *control*, *compute* and *process* tasks may be FORTRAN programs using standard I/O statements, the implementation of existing programs is much easier.

DIRECT SCF CALCULATIONS ON A TRANSPUTER NETWORK

The Hartree Fock method is probably the most important way to calculate the structures and properties of atoms, molecules and clusters ab initio, i.e. without using parameters adjusted to experimental data. Using the very sophisticated SCF program packages available today, systems containing up to 50 atoms can be treated with high accuracy.

The size of the molecules is limited by the huge quantity of two electron integrals which have to be evaluated in the course of the SCF procedure. Two variants of Hartree Fock programs exist, differing in the way they handle the two electron integrals. In the conventional version, they are calculated at the beginning of the program and stored on mass storage devices (e.g. disks). In every SCF iteration they are loaded back to the main memory in small portions to be processed to the Fock matrix elements. Limiting factors for conventional SCF calculations are the amount of available mass storage space and the I/O bandwidth.

In the direct SCF variant [15,16,17,18] the I/O problem is avoided, as the two electron integrals are evaluated in every SCF cycle. Although the computational effort increases significantly, the method is worthwhile if we use computers with a high computational / I/O performance ratio. The direct approach solely allows the treatment of large molecules. A compromise between these two variants is the semidirect SCF as implemented in the TURBOMOLE package [18]. In this method only the most expensive two electron integrals are stored, the rest is recalculated.

Using Standard Programs

Farming on the program level was investigated using the semidirect TURBOMOLE SCF part. The MultiTool environment was not considered. In this case we would have to program an optimized data flow of integrals to the disks ourselves. We did not want to undertake this development effort, which is not necessary under the Helios operating system. The implementation of the program under Helios caused only few problems, most of which were related to bugs in the Helios software, which, like every other new software product, was not at all free of bugs. To compile the 15000 lines of FORTRAN code we had to use the -Zn option to allocate more memory for the compiler. This option is mentioned nowhere in the manuals.

Table 1: Total execution times (in s) for the calculation of formic acid with a DZP basis (55 s-, p- and d-functions).

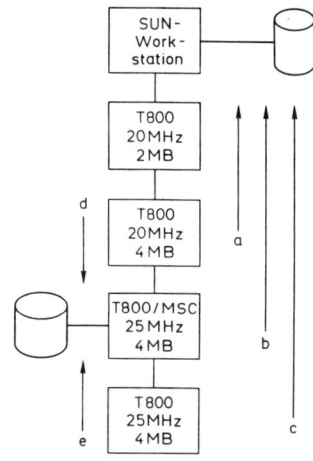

node path	single process	a+b parallel	a+b+c parallel	d+e parallel
a	3880	3930	4010	
b	3130	3170	3290	
c	3190		3250	
d	4149			4229
e	3413			3655

a) T800(20 MHz) - via 2 links - SUN disk
b) T800(25 MHz) - via 3 links - SUN disk
c) T800(25 MHz) - via 4 links - SUN disk
d) T800(20 MHz) - via 1 link - MSC SCSI disk
e) T800(25 MHz) - via 1 link - MSC SCSI disk

More serious than small software bugs is the absence of virtual memory management for the transputers. The whole executable task has to fit into the physical memory of a node. As the DSCF module needs almost 3 MB even with rather small array dimensions we could only use the nodes with 4 MB local memory.

For the test calculation we used our standard example, trans formic acid with a DZP basis set. The basis functions were taken from the TURBOMOLE basis set library. The stored part of the two electron integrals needed about 6 MB of disk space. The files were located either on the disk of the host computer (Sun workstation) or on a disk connected to the MSC node via a SCSI interface. In the latter case, we used the helios file system [14]. The results of our test are summarized in table 1 together with a diagram of the configurations used. The first column in the table shows the execution times measured without other concurrent processes. On a 25 MHz node the execution time is half of the CPU time needed on a Micro-VAX II (6074s). The node is slower than a Sun 4/280 (950s) by a factor of 3.3 .

The most interesting topic of our investigation concerned the question whether the execution time is increased significantly due to additional communication. The answer is no. Only a small difference relative to case (a) is observed, when routing the integral streams through one (b) or two (c) additional nodes. Note that the numbers for (a) and (d) have to be scaled down by a factor of 1.25 because of the lower clock rate, to be comparable to the other results.

If the same SCF program is running on the other nodes, reading and writing the two electron integrals concurrently to and from the same disk, the performance per node is decreased by only a few percent. The communication via links requires little CPU time and is controlled by the transputer very efficiently. Even if the I/O of a program is not negligible, only a part of the nodes must be equipped with mass storage devices in order to apply the farming concept on the program level. During this investigation the Helios file system for the SCSI drive attached turned out to be much slower than the access to the Sun disk.

Using A Distributed Direct SCF

A division of a direct SCF program according to the scheme in figure 3 is shown in figure 5. Vertically, the flow of the single tasks is drawn. Horizontal arrows inscribed with italic letters denote communication. For simplification the *distribute & collect* process (vid. fig.3) is omitted in figure 5. It can be regarded as a harness for the *compute* process, intervening with the communication channels. By far the most of the computing time is used by the *compute* process. Therefore, it can be replicated and distributed on a large number of processors.

The quantity of integrals calculated in the main loop of the *compute* process may be one specific integral, one batch of integrals (as shown in fig. 5) or a certain number of batches. A batch consists of all integrals derived from four shells of basis functions. A shell includes all basis functions with the same angular momentum number, the same exponent and the same location. The more integrals are calculated inside one loop cycle, the more intermediate results can be reused several times, reducing the total number of operations. For that reason, the calculation of only one integral per cycle turned out to be very inefficient (cf. [19]).

On the other hand the calculation of one or more batches per cycle may cause problems with the load balance due to the wide spread granularity of the processes. The size of a batch can differ from others by several orders of magnitude (4 different s shells: 1 Integral; 4 different d shells: 1296 integrals (cartesian gaussian functions)). The number of operations in one cycle depends not only on the l value of the basis functions, but also on the degree of contraction and the neglection algorithm for small integrals. The estimation of this number is very difficult. Therefore, a predetermined distribution of the integrals to the different nodes may result in a worse load balance [20,21]. The calculation of the two electron integrals is a very good example to show the benefits of the farming concept.

MultiTool version: In our direct SCF version running under MultiTool, the *control* and *process*

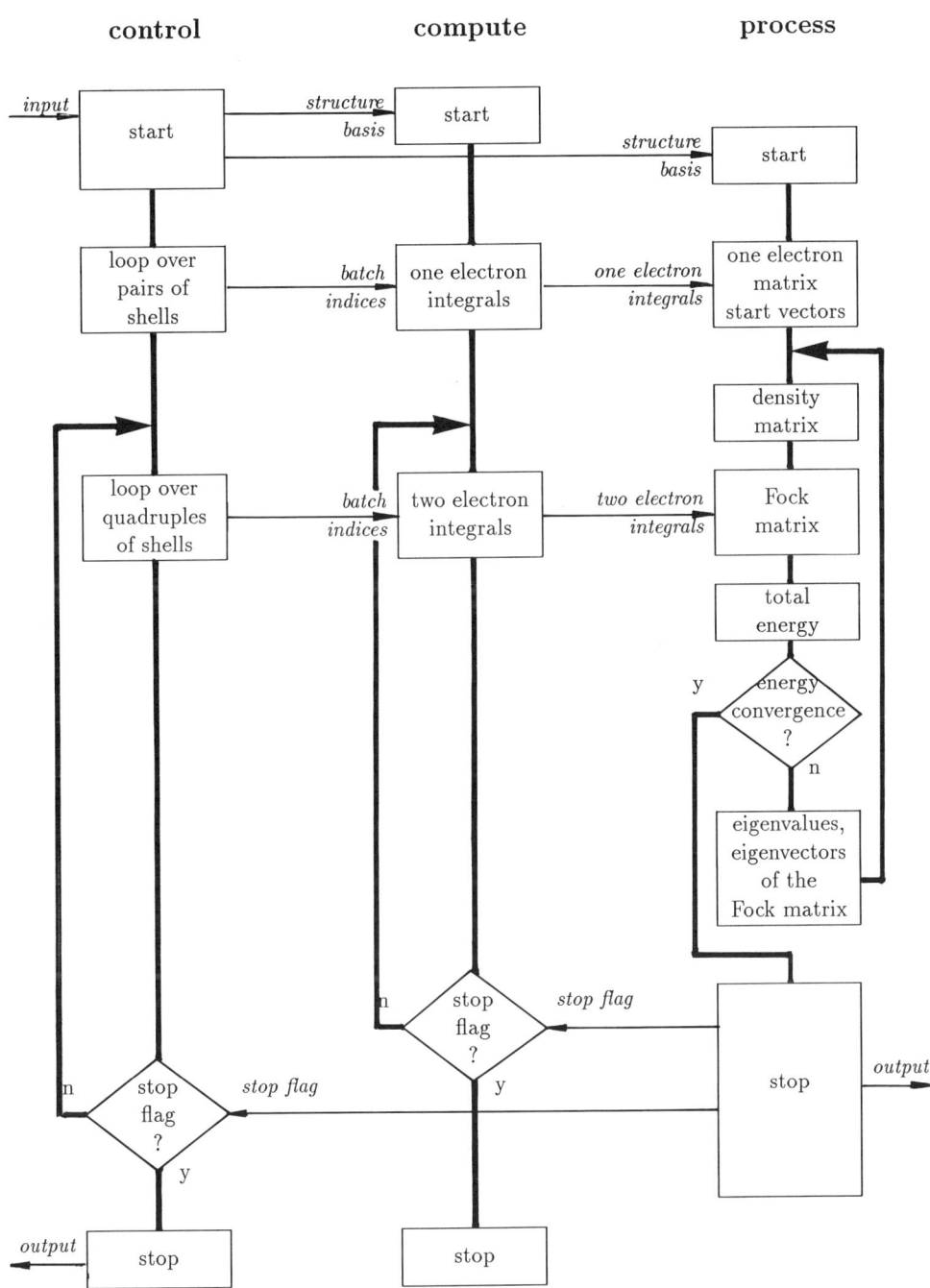

Fig. 5: Partitioning of a direct SCF program into independent tasks.

tasks (fig. 3,5) are combined in one piece of OCCAM code. The parallelism of the various parts is defined with the OCCAM PAR constructor. The distribution of the input data and the collection of the integrals is performed by distributed OCCAM processes running parallely to the *compute* tasks. They allow the configuration of linear farms or tree structures.

Details of the whole program and various test calculations are discussed in ref. [4,22]. In this article we give only a summary of our experiences.

- The most time consuming part of the Hartree-Fock method is the calculation of the two electron integrals. In contrast to the poor vectorizability of this step, it can be parallelized very efficiently. Using the farming concept, we observed a linear speed-up up to at least eight processors. Even if we consider that the matrix operations (*process* task) are performed sequentially, the speed-up of the whole program is not less than 7.1 with eight *compute* tasks.

- To be able to run the SCF program on a distributed memory parallel computer, many intermediate results have to be calculated repeatedly. With the algorithm used [23] the execution is slowed down approximately by a factor of two. This is the price for good parallelizability.

- The time to develop the distributed SCF program was shorter than expected. This was possible because we were able to embed existing FORTRAN subroutines into the OCCAM code. We used the integral subroutines from the MELD [24] program package.

The arguments pointed out in [4,22] are corroborated by further investigations. The linearity of the speed-up concerning the integral calculation has been observed also in our expanded 17 transputer network. The farming concept leads to a good load balance even if the network consists of nodes with differing performance.

A better use of intermediate results was made by treating more than one integral batch inside the main loop of the *compute* process. In fact the execution time was halved in some cases. Unfortunately, the increased granularity of the tasks involved led to load balancing problems even when using the farming mechanism. The investigations on the partitioning of the two electron integrals into larger groups of batches are not completed up to now.

In the original version of our program we used exclusively the McMurchie-Davidson algorithm [23] for the calculation of the integrals. Especially for basis functions with lower angular momentum faster algorithms are available. We now integrate step by step the subroutines developed by Bär

Table 2: Calculation of the two electron integrals using the McMurchie-Davidson algorithm (MD) [23] and some subroutines developed by Bär (B) [25]: Execution times in seconds.

B			-	$(ss\|ss)$	$(ss\|ss)$ $(ps\|ss)$	$(ss\|ss)$ $(ps\|ss)$ $(pp\|ss)$ $(ps\|ps)$
MD			all	rest	rest	rest
C_2H_4	DZ	(a)	92	77	46	27
C_2H_4	DZPD	(b)	487	465	391	302
$HCOOH$	DZP	(c)	334	325	295	252

a) 32 s- and p-functions
b) 64 s-, p- and d-functions
c) 58 s-, p- and d-functions

[25] in our integral program. These subroutines are based on the work of Obara and Saika [26] and are used in the TURBOMOLE program package [18]. The execution time improvements, measured with a linear farm of eight processors, are shown in table 2.

Helios version: In this version the *control*, *compute* and *process* tasks (fig.3) are FORTRAN programs. The only nonstandard statements are related to the inter-task communication. Instead of standard FORTRAN I/O statements we used POSIX streams, because of a bug in the interface between the FORTRAN compiler and CDL (cf. [27]). The whole program is defined by a CDL script (Example 1) using the Helios Load Balancer (lb) as the *distribute & collect* process. Up to now, the program runs only with one *compute* task, as the actions of the load balancer do not correspond with the description in the manual. A revised version of the load balancer is announced.

CONCLUSION

Future supercomputers will be parallel systems. As sure as this statement is, as uncertain is a prediction of the most promising architecture. Presently many different concepts are discussed. SIMD systems are favourable if algorithms are used for which highly vectorized code can be written. However, programs which use only a small portion of the maximum performance of a single vector processor, will waste even more computing resources on a parallel vector computer. The MIMD concept may be an expedient in this case.

The transputer is a computer on one chip, designed for parallel processing on a MIMD basis. Multidimensional networks, transfering data by *message passing* very efficiently, can be constructed with little expense. It is a favourable hardware to gain experience with parallel computing. Two different ways of writing programs for transputer networks are outlined in this article. With the distributed operating system Helios, concurrent tasks can be defined very easily by using code, written in standard programming languages. Provided that some problems emerging in the initial releases can be solved soon, Helios is a well suited software tool for transputer networks. The second way, programming in OCCAM under TDS, is more laborious. However, the programming effort can be reduced considerably by integrating existing FORTRAN code into an OCCAM harness.

The use of FORTRAN subroutines not only shortens the period of development. We also can participate in the successful investigations of other groups which nearly exclusively program their improved algorithms in FORTRAN. The insertion of the TURBOMOLE integral subroutines [18,25] is an example. We used the farming concept to get a good load balance in the network. We investigated the farming on the program and on the subroutine level. In the former case, the only limitation is the size of the local memory of a node. The routing of large data sets to remote disks in the network turned out to be no bottleneck. Farming on the subroutine level was implemented very successfully in our distributed direct SCF program. We tested the program on a network with finally 17 nodes. Up to this size a nearly linear performance increase has been observed.

The MIMD concept is very useful for SCF calculations, whereas post Hartree-Fock methods (e.g. CI) often can be vectorized very efficiently. The computing resources would be exploited best if every method is running on a computer with the most suitable architecture. Both, massively parallel MIMD-systems and SIMD-machines, should be available in very fast local or wide area networks. The distributed computing in such a large network, however, requires standardized interfaces between different programs. The definition of a standard format for structures, basis set information, matrices, orbitals, integrals, etc. is a challenge for the computational chemist in the next years.

References

[1] Saunders VR, Guest MF (1982) Comp Phys Comm 26:389

[2] Ahlrichs R, Böhm H-J, Ehrhardt C, Scharf P, Schiffer H, Lischka H, Schindler M (1985) J Comp Chem 6:200

[3] Kutzelnigg W, Schindler M, Klopper W, Koch S, Meier U, Wallmeier H (1986) In: Dupuis M (ed) Supercomputer simulations in chemistry. Springer, Berlin Heidelberg New York

[4] Wedig U, Burkhardt A, von Schnering H-G (1989) Z Phys D 13:377

[5] Buell DA et al (1988) J Supercomputing 1:301

[6] INMOS Ltd (1987) The transputer family 1987. INMOS, Bristol

[7] INMOS Ltd (1988) occam 2 reference manual. Prentice Hall, London

[8] Hey AJG (1988) Comp Phys Comm 50:23

[9] Hey AJG (1989) Comp Phys Comm 56:1

[10] Glendinning I, Hey A (1987) Comp Phys Comm 45:367

[11] Transputer Development System (D700 D). INMOS, Bristol

[12] MultiTool 5.0. parsytec, Aachen

[13] Perihelion Software Ltd (1989) The helios operating system. Prentice Hall, London

[14] Helios file system, version 1.1. parsytec, Aachen

[15] Almlöf J, Faegri Jr. K, Korsell K (1982) J Comp Chem 3:385

[16] Almlöf J, Taylor PR (1984) In Dykstra CE (ed) Advanced theories and computational approaches to the electronic structure of molecules. Reidel, Dordrecht

[17] Cremer D, Gauss J (1986) J Comp Chem 7:274

[18] Häser M, Ahlrichs R (1989) J Comp Chem 10:104

[19] Colvin ME (1986) Ph. D. thesis LBL–23578. Univ California, Berkeley

[20] Dupuis M, Watts JD (1987) Theor Chim Acta 71:91.

[21] Guest MF, Harrison RJ, van Lenthe JH, van Corler LCH (1987) Theor Chim Acta 71:117

[22] Wedig U, Burkhardt A, von Schnering HG (1990) In: Grebe R (ed) Parallele Datenverarbeitung mit dem Transputer. Springer, Berlin Heidelberg New York

[23] McMurchie LE, Davidson ER (1978) J Comp Phys 26:218

[24] McMurchie L, Elbert ST, Langhoff SR, Davidson ER et al; Program MELD. Univ Washington, Seattle

[25] Bär M (1988) diploma thesis. Univ Karlsruhe

[26] Obara S, Saika A (1986) J Chem Phys 84:3963

[27] Allan R (1989) Parallelogram 2:21:17

Parallelizing an SCF program on SUPRENUM

Ulrich Meier
Ruhr University Bochum, Lehrstuhl für Theoretische Chemie, 4630 Bochum, Universitätsstr. 150

Reiner Vogelsang
SUPRENUM GmbH, Hohe Str. 73, D-5300 Bonn 1

Abstract

This article describes how an existing serial SCF-Programm can be adpated to a parallel computer architecture with distributed memory like the SUPRENUM system. The SCF program, which should be adapted, can be considered as a traditionally coded Gaussian lobe program as far as the code structure is concerned. The intregral part of the program is massively vectorized.

A short overview of the SUPRENUM system describes the basic features of the hardware, programming environment, and the user interface.

1 The SUPRENUM system

1.1 The hardware

Figure 1 shows the overall structure of the SUPRENUM hardware [1]. In the SUPRENUM system 256 nodes are connected via a two-level interconnection network of buses:

- 16 *clusters* are connected via a torus like bit serial bus network (SUPRENUMBUS) providing 4 links per cluster.

- Within a cluster 16 *nodes*, a cluster disk controller able to handle ut to 4.8 Gbytes of disk space, a diagnosis node, and an communication node driving the SUPRENUMBUS links are connected via a 64 bit parallel bus (*Clusterbus*, 320 Mbytes/s)

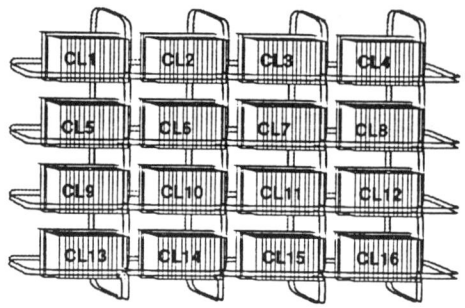

SUPRENUM Bus:

Parallel Token Ring
Transfer rate per Link 12.5 Mbyte/s
4 Links / Cluster

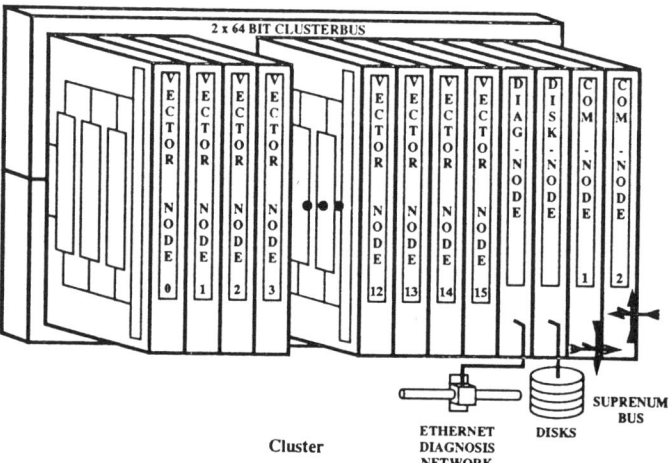

Figure 1: Structure of the SUPRENUM-prototype with 256 nodes in 16 clusters (top) and the inner structure of a cluster

Each *node* (see figure 2) consists of the MC68020 CPU plus coprocessor, 8 Mbyte of private memory with EDC (error correction and detection), a fast floating-point vector unit (10 Mflops peak performance, 20 Mflops with chaining), and dedicated communication hardware. The communication hardware and the VFPU are handled like coprocessors by the CPU, i.e. they can run independently from each other.

The hardware described so far is called the SUPRENUM-kernel, which is accessed via a gateway . The gateway can be connected to a LAN of workstations or other machines.

The SUPRENUM system is a scalable architecture. The performance can be easily enhanced by adding more clusters or nodes to an existing system. Therefore, the SUPRENUM system covers a peak performance range from 0.32 GFLOPS (1 Cluster) ut to 5 GFLOPS (16 Clusters).

Migrating from a one-cluster system to a multi-cluster system is completely transparent to the user since the system software (see section 1.2) supports a software concept based on processes (for instance multitasking on each node) and message passing.

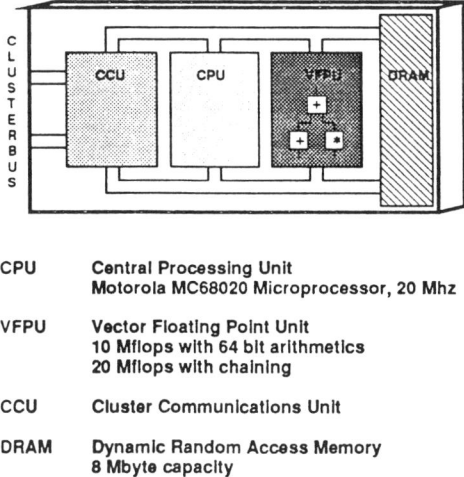

CPU	Central Processing Unit
	Motorola MC68020 Microprocessor, 20 Mhz
VFPU	Vector Floating Point Unit
	10 Mflops with 64 bit arithmetics
	20 Mflops with chaining
CCU	Cluster Communications Unit
DRAM	Dynamic Random Access Memory
	8 Mbyte capacity

Figure 2: Hardware of a node with CPU kernel, VFPU and communication hardware.

1.2 The system software

The software concept for SUPRENUM is based on a *process* system and on *message passing communication* handling. The process concept (the so-called *abstract SUPRENUM architecture*) is a dynamic one and is characterized by the following elements:

- Processes (or tasks) are autonomous program units which run in parallel.
- Processes can terminate themselves and can create, but not terminate other processes.
- Processes communicate only by exchange of messages, and no shared memory is available.
- Applications are started by one initial task.
- In arithmetic expressions and communication instructions array constructs are especially supported.
- The user defined process system is homogeneous and independent from the actual hardware configuration. The two-level architecture (cluster structure) is not reflected in the abstract SUPRENUM architecture and is completely transparent to the user. The processes are mapped to the clusters and nodes at run-time.

Figure 3 shows that this abstract SUPRENUM architecture is the central model in the system software. The user should write his or her codes only in terms of processes.

The mapping of processes to nodes is supported by the *mapping-library* [4]. It provides optimal mapping strategies for some standard process systems (such as trees, rings, grids) and uses heuristical strategies for other process structures.

The SUPRENUM *operating sytem* consists of three components residing on the front end system, the cluster level and on the node level. The front end system is operated under UNIX V[1]. On the cluster level the operating system supports the local disk, the performance analysis, and the connection between the two communication levels. In each node a small operating system (PEACE) [9] is responsible for the process scheduling and the message handling.

The *programming language* for numerical computations is FORTRAN [2], an extended FORTRAN 77. The extensions include special process handling and message-passing constructs and an array syntax formulation according to the proposed FORTRAN-8X standard (see figure 4). A state-of-the-art *auto-vectorizer* will be available which automatically generates code for the vector nodes – if the user does not want to use the new Fortran-8X array notation. This allows porting of existing "dusty deck" software since large parts of the code can be kept without modifications [13]. Later on, a semi-automtatic parallelizer will be available which will support

---
[1] registered trademark of AT&T

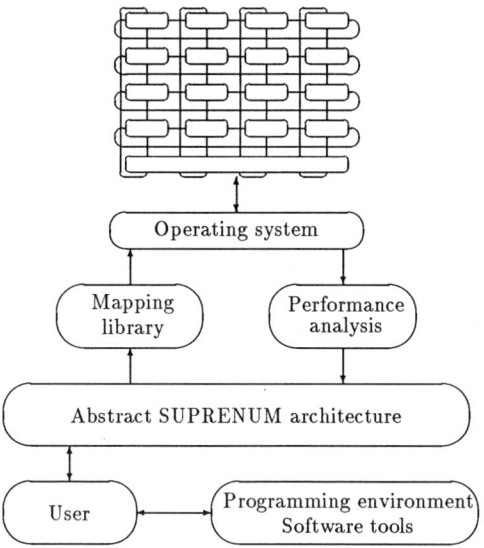

Figure 3: The SUPRENUM system software

the user in the construction of host and node processes and which automatically distributes arrays over the nodes [12].

In addition a parallel version of C will be available.

A *communications* library for grid applications allows easy, safe, and portable programming for regular and block-structured grids [3].

The application software based on the abstract SUPRENUM machine can be developed (in FORTRAN-8X) by using *simulators* [5]. Thus nearly the complete program development is done on workstations where the simulator checks the logical correctness of the communication and gives performance estimates. The only step in the program development where the parallel multiprocessor system is really needed is the final tuning.

The communication between and the synchronization of processes is done via message-passing. SUPRENUM offers an *asynchronous message-passing* model, i.e. the sending application program does not have to wait until the message has arrived at the receiver and can continue immediately. On the receiving side, the message arrives in the mailbox and can be selected by the application (see figure 5).

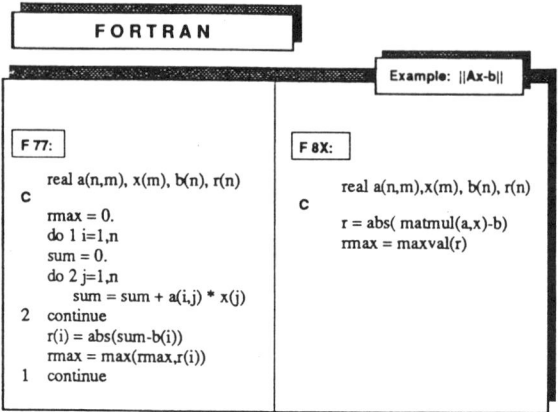

Figure 4: Comparison between FORTRAN-8X and FORTRAN 77 for a simple example

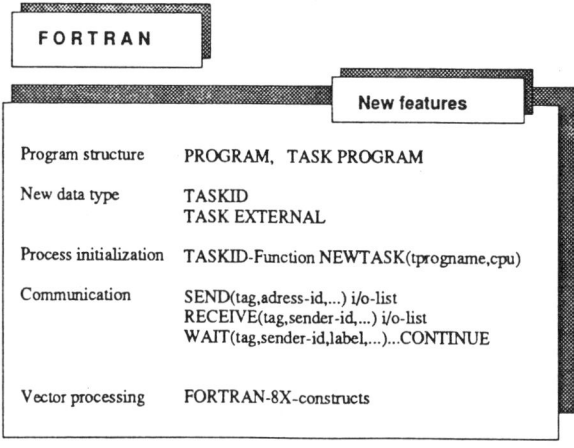

Figure 5: Basic statements for message passing and process creation

2 Parallelization of an SCF program

During the last years the computer aided molecular design has become more and more important. The reason for this trend is, for instance, the need to reduce the time for the development of new chemical compounds. The methods of computer aided design (molecular dynamics, 'ab initio' and semi-empirical methods of quantum chemistry) can give in interaction with the experiment answers for certain problems occuring during the development of chemical compounds.

For molecules of interest for the experimental working chemists the methods of 'ab initio' quantum chemistry requires supercomputer facilities. Certain time intensiv program parts of this class of applications are well suited for machines with distributed memory like the SUPRENUM system.

The program IGLO considered whithin the following sections of this paper is an MO-LCAO-SCF-program with a special part based on the coupled Hartree Fock theory to calculate magnetic properties of molecules [21,16,17,18]. As far as the adaptation of this program to the SUPRENUM system is concerned, only the SCF-part of IGLO will be discussed within the scope of this paper.

2.1 The LCAO-MO-SCF-algorithm

As the details of the LCAO-MO-SCF-algorithm can be looked up in scientific publications and review articles (see [14] and the references there, [15,19,20]), we are reporting for the closed shell case the basic features of the algorithm only.

The time independent Schroedinger equation

$$\hat{H}\Psi(\vec{r},\vec{R}) = E(R)\Psi(\vec{r},\vec{R})$$

$$\hat{H} = \underbrace{-\frac{1}{2}\sum_{i=1}^{n}\Delta_i - \sum_{i=1}^{n}\sum_{S=1}^{M}\frac{Z_S}{r_{iS}}}_{h(1)} + \underbrace{\sum_{i<j}^{n}\sum\frac{1}{r_{ij}}}_{V(12)}$$

can be approximately solved by using a one-Slater-determinant-Ansatz for Ψ:

$$\Psi \approx \Phi_0 = (n!)^{1/2}\sum_P(-1)^P[\psi_1(1)\psi_2(2)\ldots\psi_{n/2}(n)]$$

The molecular orbitals (MOs) ψ are then approximated as a linear combination of basis functions ϕ (LCAO):

$$\psi_i = \sum_\mu^N c_{\mu i}\phi_\mu$$

The basis functions are then represented by an expansion with simpler functions, for instance Gaussian lobes:

$$\phi_\mu = \sum_s d_{\mu s} g_s$$

$$g_s(\alpha, r) = \left(\frac{2\alpha}{\pi}\right)^{3/4} exp(-\alpha r^2)$$

The coefficients $c_{i\mu}$ are then determined by variation:

$$E = \left(\phi_0|\hat{H}|\phi_0\right)$$

$$\delta E \stackrel{!}{=} 0$$

The condition above leads to the Rothaan-Hall-equation for the determination of the coefficients $c_{\mu i}$:

$$\left(\underline{\underline{F}} - \varepsilon \underline{\underline{S}}\right)\underline{\underline{C}} = 0$$

$$\underline{\underline{F}} = \underline{\underline{h}} + \underline{\underline{G}}$$

with

$$h_{ij} = \left(\psi_i|\hat{h}(1)|\psi_j\right)$$
$$S_{ij} = (\psi_i|\psi_j)$$
$$G_{ij} = \sum\sum_{k\leq l} R_{kl}\left[4(ij|kl) - (ik|jl) - (il|jk)\right]$$

where

$$(ij|kl) = \int d\vec{r}_1\vec{r}_2 \psi_i(1)\psi_j(1)\frac{1}{r_{12}}\psi_k(2)\psi_l(2)$$

are the two electron integrals, and

$$R_{kl} = \sum_\mu c_{\mu k}c_{\nu l}(1-\delta_{kl})$$

is the density matrix. Since the Fock operator $\underline{\underline{F}}$ depends on $\underline{\underline{C}}$ one has to iterate until self consistency is reached.

In a algorithmic notation the following steps are then performed in an SCF-program:

1st Initial guess of the vectors $\underline{\underline{C}}$

2nd Calculate one ($\sim N^2$) and two electron integrals ($\sim N^4$)

3rd Build $\underline{\underline{R}}$ and $\underline{\underline{F}}$

4th Solve pseudo eigenvalue problem $\left(\underline{\underline{F}} - \varepsilon \underline{\underline{S}}\right) \underline{C} = 0$

5th Calculate Energy $E^{(n)} = 2Tr(\underline{\underline{R}}\,\underline{\underline{F}}) - Tr(\underline{\underline{R}}\,\underline{\underline{G}})$

6th $E^{(n)} - E^{(n+1)} \stackrel{?}{=} 0$, go back to 2nd

7th Calculation of properties

It is obvious that the time consuming step 2 - a problem of four indices - is the main target for optimizations, i.e. parallelization, since the two electron integrals can be calculated independently from each other. The other steps of an SCF program are of order N^3 or N^2.

2.2 Code analysis of IGLO and parallelization concept

The main structure of IGLO can be regarded as traditional. The two electron integrals are calculated before the SCF iteration and written on disk. This strategy may hold for small molecules. For large molecules there is no way around the direct scheme of an SCF program, i.e. recalculation of the two electron integrals for each SCF iteration, since disk capacity and disk access time can be a severe bottleneck. However, for large production runs IGLO was adapted to the CYBER 205. For this adaptation the disk i/o was optimized by using concurrent buffer routines working on 4kword blocks and the two electron part was massively vectorized. The vectorization was done over groups (for details, see [22]). In order to achieve high performance a blocked vectorisation scheme was invented:

> For the four folded loop I, J, K, L of the integral program lower bounds, IMIN, and upper bounds, IMAX, are determined for the outer loop, so that the integrals can be calculated in large blocks of a predefined vector length in the kernel of the integral program. The bounds of the inner loops and the disk i/o are controlled by IMAX, too.

This vectorization scheme can be directly exploited for a straight forward parallelization of the integral program:

1st A master creates tasks (slaves) built by the modules of the integral program and supplies each process with an IMIN and IMAX in such a way that a good load balance is achieved. The initialization of the tasks and the determination of the loop bounds, IMIN and IMAX, is done before the start of the SCF iterations.

2nd During the SCF iterations the master receives asyncronously from its slaves the two electron integrals. The calculation of the integrals can be overlapped with the computations in the iteration loop (the diagonalization or building of the Fock matrix, for instance).

Apart from having a nearly linear spead up of the computation of the two electron integrals, no disk space is necessary any more.

The question is how this concept can be realized in practice for a code developed over several years.

2.3 Programming aspects

The modules for the computations of two electron integrals are building nearly a stand alone program unit within IGLO (for the main call sequence of this modules see figure 6). Apart from two subroutines, they can be cut from the original code without any modifications building the so called task program unit to be run in parallel on the hardware. In the original code the

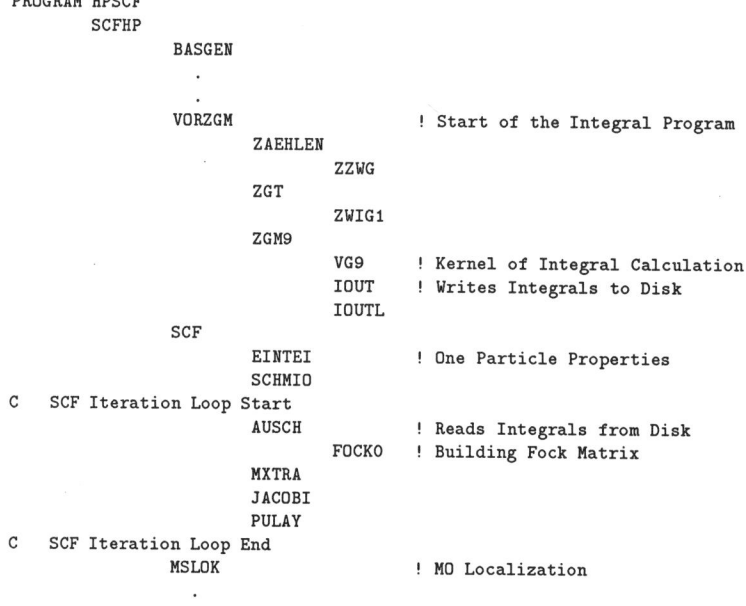

```
PROGRAM HPSCF
       SCFHP
              BASGEN
                  .
              VORZGM                   ! Start of the Integral Program
                     ZAEHLEN
                            ZZWG
                     ZGT
                            ZWIG1
                     ZGM9
                            VG9        ! Kernel of Integral Calculation
                            IOUT       ! Writes Integrals to Disk
                            IOUTL
              SCF
                     EINTEI            ! One Particle Properties
                     SCHMIO
C     SCF Iteration Loop Start
                     AUSCH             ! Reads Integrals from Disk
                            FOCKO      ! Building Fock Matrix
                     MXTRA
                     JACOBI
                     PULAY
C     SCF Iteration Loop End
              MSLOK                    ! MO Localization
                  .
                  .
```

Figure 6: Main call sequence of IGLO

call to the two electron integral program (VORZGM) were substituted by a new subroutine SDINIT which creates the processes and sends initializing data to them. This data are received in the task program by a new subroutine RCINIT (see appendix A). The interface between the SCF iterations and the two electron program is provided in the serial version of IGLO by two io routines (AUSCH and IOUTL). The io routines are used in the parallelized version as an interface, too, by substituting the read/write statement by RECEIVE and SEND (see appendix B).

The final structure of the parallelized version of IGLO the displayed in figure 7.

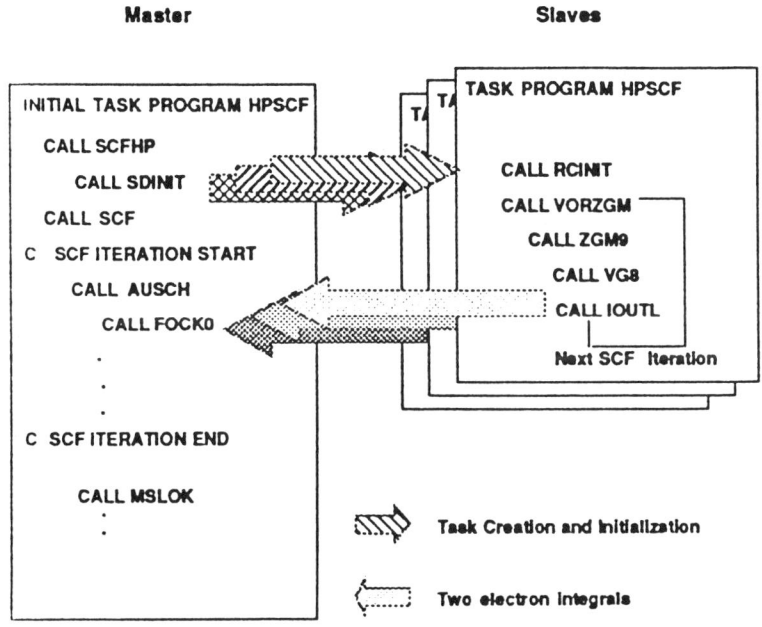

Figure 7: Parallelized version of IGLO with communication structure

3 Current state

The parallelized program is still in a tuning phase since several points can be optimized:

The *vectorization* can be regarded 'CYBER 205'-like, i.e. memory bound. Here some vectorizable loops have to be restructured to support the vector register model of the VFPU of the SUPRENUM node. Basicly, this means a removal of redundant load and store operations from the code.

As the number of nodes increases the sampling of the intgrals in the current version can lead to unnecessary long wait states of messages in the mail box of the master. Therefore, the *communication* strategy has to be optimized by implementing a binary tree or spanning tree sampling scheme.

Further parallelization has to be done. Rembering Amdahl's law - originally deduced for vector maschines - the efficient pallelization of a time consuming step in a large program is only a portion of the truth. Other steps, the two index transformation or diagonalization can become a time determining step as the number of nodes increases with constant problem size. For instance, the eigenvalue calculation can be substituted by a parallel version of the SLAP-library (SUPRENUM Linear Algebra Package).

The final execution times of the parallelized version of IGLO will be published in the near future.

4 Conclusions

A rather complex code like a quantum chemical program can be ported to an distributed memory machine like the SUPRENUM system without major changes of computational kernels. The parallelization of IGLO is supported by an advanced and user friendly communication interface. We hope to adapt other program packages of interest for computational chemistry in the next future and are open to further discussions on this topic.

5 Appendix A

Figure 8 shows the task program unit calling the subroutine VORZGM. The subroutine VORZGM and all subroutines called by VORZGM were cut from the serial version of IGLO. N replicates of this task program units are created by the subroutine SDINIT (see figure 9). The subroutine SDINIT is called in IGLO instead of VORZGM. SDINIT sends inital data and task specific data to the tasks indentified by their task ids. The data are received by the subroutine RCINIT. The task program shows also the use of the WAIT statement for program flow control via message passing. The WAIT statement can be used for the implementation of a socalled farming concept (see for conceptional aspects [23,24]).

Figure 8: The task program unit HPVORZGM calling the integral computation

```
      task program hpvorzgm
cf****************************************************************
      parameter (lpp = 3)
      parameter (lpg = 65536)
      parameter (nr = 4093)
c
      implicit double precision (o-z, a-g)
      integer go_ahead,stopper
c
      common /page/ tr1(nr), tr2(nr), tr3(nr), nit, hijklt(nr)
     &, hr1(nr), hr2(nr), hr3(nr), nin, hijkl(nr), hu(2)
c
      common // a(lpp*lpg)

c     Settings of control tags

c     If a message is in the mailbox with tag start_scf, start
c     integral computation
      go_ahead=9997
c     If a message is in the mailbox with tag stoper, terminate
c     the integral computation!
      stopper=9999

      call recinit

c     Alternative wait
 100  wait(tag=go_ahead, continue, tag=stopper, label=9999)
      call vorzgm(a,lpp)
      goto 100

c     Termination of the application
 9999 continue
      write(6,*) ' Process HPVORZGM Nr.',cpuid,' terminates!'
      end
```

Figure 9:

```
      subroutine sdinit()
c************************************************************************
c*    Creates and initializes the tasks calculating the two electron    *
c*    integrals.                                                        *
c************************************************************************
      parameter (ngr = 255, nlo = 750)
      parameter (mxnodes = 256)
c
      implicit double precision (o-z, a-g)
c
      common /bn/ n1, idoof(31)
      common /bii/ igi(ngr + 3)
      common /gd/ ak(nlo)
      common /bk/ bh(4, nlo)
      common /eta/ bhx(nlo), bhy(nlo), bhz(nlo), bhe(nlo), kind(nlo + 1)
      common /grr/ nnr, nnr64, nrec, nrec64, lcpa, lcpa64, ithr, thr,
     &tr64, nsym, io64, nlhf, ngsy, imxx, iix(2)
      common /gusy/ iel, nhrl, na(ngr, 8), dsy(3, 64)
      common /page1/ m7, isym(6, ngr)
      common /errt/ iimax,ixum,dix,ff1(1700),ff2(1700),ff3(1700),
     &ff4(1700)
csuprenum
      integer imin0     (maxnodes)
      integer imax0     (maxnodes)
      integer nrecloc   (maxnodes)
      integer nrec64loc(maxnodes)
      integer nnrloc    (maxnodes)
      integer nnr64loc (maxnodes)
      integer cpuid(maxnodes)
      taskid pid(maxnodes)
      common /commun/ imin0,imax0,nrecloc,nrec64loc,nnrloc,nnr64loc
     &                ,mynode,nnodes,pid,cpuid

c     Name of the task program unit
      task external hpvorzgm
csuprenum


c     Determination of IMIN and IMAX for each node
      call dryrun (nnodes,n1,imin0,imax0)

c     Creation of tasks

c     Mapping of the Processes on the CPUs
      do i=1,nnodes
        cpuid(i)=i
      enddo

c     Multiload
      pid(1:nnode)=newtask(hpvorzgm,cpuid(1:nnode))

c     End of creation

c     Sending of global data
```

```
      send(taskid=all,tag=1) n1,m7,nsym,lcpa,lcpa64,ithr,nsym,io64
     &                      ,nlhf,ngsy,imxx,iix(2),iimax,ixum,igi(n1+3)
     &                      ,na(1:n1,1:8),iix(1:2),isym(1:6,1:n1)
     &                      ,kind(m7+1),thr,thr64,dix,ak(1:m7)
     &                      ,bh(1:4,1:m7),bhx(1:m7),bhy(1:m7),bhz(1:m7)
     &                      ,bhe(1:m7),dsy(3,64),ff1(1700),ff2(1700)
     &                      ,ff3(1700)

c     Sending node specific data
      do i=1,nnodes
         send(taskid=pid(i),tag=1000))imin0(i),imax0(i),cpuid(i)
     &                      ,nnodes,pid(1:nnodes)
      enddo

      return
      end
```

Figure 10:

```
      subroutine rcinit()
c***********************************************************************
c*    Receives initial data from the initial task                      *
c***********************************************************************
      parameter (ngr = 255, nlo = 750)
      parameter (mxnodes = 256)
c
      implicit double precision (o-z, a-g)
c
      common /bn/ n1, idoof(31)
      common /bii/ igi(ngr + 3)
      common /gd/ ak(nlo)
      common /bk/ bh(4, nlo)
      common /eta/ bhx(nlo), bhy(nlo), bhz(nlo), bhe(nlo), kind(nlo + 1)
      common /grr/ nnr, nnr64, nrec, nrec64, lcpa, lcpa64, ithr, thr,
     &tr64, nsym, io64, nlhf, ngsy, imxx, iix(2)
      common /gusy/ iel, nhrl, na(ngr, 8), dsy(3, 64)
      common /page1/ m7, isym(6, ngr)
      common /errt/ iimax,ixum,dix,ff1(1700),ff2(1700),ff3(1700),
     &ff4(1700)

csuprenum
      integer nnodes,imin0,imax0,cpuid
      taskid pid(maxnodes)
c     Function master returns the taskid of the intial task.
      taskid master
      common /commun/ imin0,imax0,nnodes,pid,cpuid
Csuprenum


c     Receiving of intial data
      receive(taskid=master(),tag=1) n1,m7,nsym,lcpa,lcpa64,ithr,nsym,io64
     &                      ,nlhf,ngsy,imxx,iix(2),iimax,ixum,igi(n1+3)
     &                      ,na(1:n1,1:8),iix(1:2),isym(1:6,1:n1)
     &                      ,kind(m7+1),thr,thr64,dix,ak(1:m7)
```

```
     &                   ,bh(1:4,1:m7),bhx(1:m7),bhy(1:m7),bhz(1:m7)
     &                   ,bhe(1:m7),dsy(3,64),ff1(1700),ff2(1700)
     &                   ,ff3(1700)

c     Receiving node specific data
      receive(taskid=master(),tag=2)imin0,imax0,cpuid
     &                              ,nnodes,pid(1:nnodes)

      return
      end
```

6 Appendix B

Figure 11 shows part of the io-routine IOUT of IGLO with the modifications needed to run in the parallel version of IGLO. The WRITE statements are substituted by SEND statements. Figure 12 shows part of the corresponding i/o routine AUSCH. The READ statements are substituted by RECEIVE statements.

Figure 11: The subroutine IOUT modified to send part of the integrals to the initial task

```
      subroutine iout(lu, lu64, nn, thre, tr64, lzza, i, r1, r2, r3, f,
     &ijklv, idbi, qq1, qq2, qq3, qq4, qq5, hijkl64, hin, hr1z, lp, lvg
     &, nl, mst, ibs, isz)
c*******************************************************************
      implicit double precision (o-z, a-g)
      parameter (nr = 4093, nx = (4 * nr) + 2)
csup
      integer stopper,go_ahead,finished
csup
      logical io64
      common /page/ tr1(nr), tr2(nr), tr3(nr), nit, hijklt(nr)
     &, hr1(nr), hr2(nr), hr3(nr), nin, hijkl(nr)
      dimension r1(*), r2(*), r3(*), f(*), ijklv(*), qq1(*), qq2(*), qq3
     &(*), qq4(*), qq5(*), hijkl64(*), hin(*), hr1z(*), idbi(*)
      common/commun/.......,cpuid
                                     .
                                     .
csup
      go_ahead=9997
      finished=9998
      stopper =9999
      wait(tag=gohead,continue)
csup
      do 100 j=1,n

                                     .
                                     .
      if (io64) then
        if (hqq .gt. htr64) then
          nit = nit + 1
          nnr64 = nnr64 + 1
```

```
              tr1(nit) = r1(j) * ff
              tr2(nit) = r2(j) * ff
              tr3(nit) = r3(j) * ff
              hijklt(nit) = ijklv(j)
              if (nit .eq. nr) then
                 nrec64 = nrec64 + 1
CSUP                write(unit=lu64) tr1
CSUP                write(unit=lu64) tr2
CSUP                write(unit=lu64) tr3
CSUP                write(unit=lu64) nit, hijklt
                 send(taskid=master(),tag=1064)tr1,tr2,tr3, nit, hijklt
                 nit = 0
              endif
              goto 99
           endif
        endif

csup  Send the final state of the integral computation to the initial task.
      if(nnr.eq.0.or.nnr64.eq.0)
         send(taskid=master(),tag=stopper)
         stop
      else
         send(taskid=master(),tag=finished) cpuid,nnr,nnr64
      endif
csup
      return
      end
```

Figure 12: The subroutine AUSCH modified to pick up the integrals calculated by the task programs.

```
      subroutine ausch(ityp, a, b, izahl, rms)
c************************************************************************
              call fock0 (a, b)

      return
      end
      subroutine fock0 (a, b)
c************************************************************************
      implicit double precision (o-z, a-g)
      implicit real (h)
      integer ready_to_receive
      parameter(maxnodes=256)
      parameter (lr = 4093, lh = 8192, lulim = 3750)
      dimension a(*), b(*)
      common /page/ hr1(lr),hr2(lr),hr3(lr),ma,ijkl(lr),
     &              r1(lr), r2(lr), r3(lr)
      common /commun/....nnrloc(maxnodes),nnrloc64(maxnodes)...
     &              ,nnodes,..,cpuid
```

```
              inode=0
              lu = 13
      csup    rewind (unit=lu)
      csup    do 2000 irec = 1, nrec
      CSUP    read(unit=lu,err=1901,end=1951) hr1
      CSUP    read(unit=lu,err=1901,end=1951) hr2
      CSUP    read(unit=lu,err=1901,end=1951) hr3
      CSUP    read(unit=lu,err=1901,end=1951) ma, ijkl

      csup    Send ready state signal to all tasks
              ready_to_receive=9997
              send(tag=ready_to receive,taskid=all)

      csup    Message for 32-bit properties
       2000   wait(tag=1032,continue

      csup    Message for 64-bit properties
           &       ,tag=1064,label=2001

      csup    Signal that one task has finished the integral computation
           &       ,tag=9998,label=998

      csup    Error signal
           &       ,tag=9999,label=999)

              receive(tag=1032) hr1,hr2,hr3,ma,ijkl
              nnr=nnr+ma
      csup
              do 3000 m=1,ma
              rx = hr1(m)
              ry = hr2(m)
              rz = hr3(m)
      c
      c       unpacking of labels   ijkl
      c
              include 'auspack.h'
              ii = itab(i)
              jj = itab(j)
              kk = itab(k)

       3000 continue
       2001 continue

      csup
              goto 2000
      csup    Stopper bug
        999 stop

      csup    Are all integrals received and is fock operator built?
        998 inode=inode+1
              receive(finished)cpuid,innr,innr64
              nnrloc(cpuid)=nnrloc(cpuid)+innr
              nnrloc64(cpuid)=nnrloc64(cpuid)+innr64
              if(inode.lt.nnodes) goto 2000
              return
              end
```

References

[1] **Giloi, W.K.:** SUPRENUM – a trendsetter in modern supercomputer development. In [11].

[2] **Bolduc, R.:** SUPRENUM Fortran Syntax Specifactions. SUPRENUM Report 7, SUPRENUM GmbH, Bonn, 1987.

[3] **Hempel, R.:** The SUPRENUM communications subroutine library for grid-oriented problems. Report ANL-87-23, Argonne National Laboratory, 1987.

[4] **Krämer, O.:** SUPRENUM – Mapping Library, User Manual. Report, GMD, St. Augustin, 1987.

[5] **Limburger, F., Scheidler, Ch., Tietz, Ch., Wessels, A.:** Benutzeranleitung des SUPRENUM-Simulationssystems SUSI. GMD, St. Augustin, 1986.

[6] **Mierendorff, H., Kolp, O., Görg, B., Lieblang, E.:** Leistungsuntersuchungen von parallelen Algorithmen für einfache Mehrgitterverfahren auf SUPRENUM-Systemen, GMD-Studien, St.Augustin, to appear.

[7] **Mierendorff, H., Trottenberg, U.:** Performance evaluation for SUPRENUM systems. Proc. UNICOM Seminar on Evaluating Supercomputers, 1-3 June 1988, London.

[8] **Peinze, K., Thole, C.A., Thomas, B., Werner, K.H.:** The SUPRENUM Prototyping Programme. SUPRENUM Report 5, SUPRENUM GmbH, Bonn, 1987.

[9] **Schröder, W.:** PEACE: The distributed SUPRENUM operating system. In [11].

[10] **Solchenbach, K.:** Grid applications on distributed memory architectures: Implementation and evaluation. In [11].

[11] **Trottenberg, U. (ed.):** Proceedings of the 2nd International SUPRENUM Colloqium "Supercomputing based on parallel computer architectures". Parallel Computing 7, North Holland, 1988.

[12] **Zima, H.P., Bast, H.-J., Gerndt, H.M.:** SUPERB: A tool for semi-automatic MIMD/SIMD parallelization. Parallel Computing 6, pp-1-18, North Holland, Amsterdam, 1988.

[13] **Ashauer R., Hoppe Th., Jost G., Krause M., Solchenbach M.** The SUPRENUM FORTRAN Compiler - Architecture and Performance, appearing in SUPERCOMPUTERS, 1990.

[14] **Hehre Warren J., Radom Leo, Schleyer Paul v. R., Pople John A.:** Ab Initio Molecular Orbital Theory, John Wiley & Sons, 1986.

[15] **Levine I. N.** Quantum Chemistry, 3rd ed., Allyn and Bacon, Boston, 1983.

[16] **Stevens R. M., Pitzer R. M., Lipscomb W. N.** J. Chem. Phys. 38,550 (1963).

[17] **Stevens R. M., Pitzer R. M., Lipscomb W. N.** J. Chem. Phys. 38,550 (1963).

[18] **Lipscomb W. N.** Adv. Mag. Resonance, Vol. 2, p. 137, 1966.

[19] **Roothaan C. C. J.** Rev. Mod. Phys. 23,69 (1951).

[20] **Hall G. G.** Proc. Roy. Soc. (London), A205,541 (1951).

[21] **Schindler M.** Phd Thesis, University of Bochum, 1980.

[22] **Meier U.** Diploma Thesis, University of Bochum, 1985.

[23] **Wedig U., Burkhardt A., Schnering H.G. v.** Z. Phys. D 13,377 (1989).

[24] **Guest M., Lenthe J. H. van** Theory and Computational Science, Appendix to the Daresbury Anual Report 1988/89,1989 ISSN: 0265-1831.

Drug Design: A Combination of Experiment and Computational Chemistry

H. P. Weber
SANDOZ A. G., Preclinical Research, CH-4002 Basel

A "Rational Drug Design" depends on the explicit formulation of a Stucture-Activity-Hypothesis which postulates directly a relation between specific molecular parameters and the biological activity of a series of drug molecules as measured in a particular biological/pharmacological assay. The molecular parameters used in these relations are usually structural dimensions (e.g. distance between functional groups defining a pharmacophor, or defining a fitting molecular shape), and electronic parameters (e.g. particular electron density distributions, location of HOMO/LUMO molecular orbitals, etc.).

Most of these parameters can be both determined experimentally (e.g. by XRay, NMR and chemical methods) and theoretically. The purpose of a Rational Drug Design must be to predict new active molecules on the basis of theoretical models. However, the models have to be derived initially from experimental data in order to be realistic.

The combined application of (limited) experimental data and quantum chemical methods in Drug Design shall be illustrated by two practical examples:

(1) In the course of the investigation of "DNA minor groove binding" molecules, the question of the directionality of the H-bond to the charged oxygens of the DNA-phosphate group emmerged. A statistical analysis of the experimental data about the H-bond directionality in phosphate-diesters from the Cambridge

Crystal Data Base showed conclusively that the P - O ... H angle clusters around 120 , and indicated vaguely a preference of the O - P - O ... H torsion angle around 0 and 180 .

```
       \
        C - O     O ... H - O
             \   /           \
              P (-)           H
             /   \
        C - O     O
       /
```

The latter hypothesis was tested by a theoretical calculation by a GAUSSIAN86 MO energy optimisation with a 6-311G** basis on a phosphate ... water model system with various torsion angles. The results did not support the hypothesis.

(2) The X-ray structural data of the complex of the enzyme chymotrypsin and the inhibitor phenethyl-boronic acid suggested an interaction between the boron and the "active site" serine oxygen. To investigate the nature of this interaction, i.e. ionic versus covalent, a quantum chemical calculation on the model system:

```
          HO                      H
            \                    /
       Me -  B . . . . . . O
            /                    \
          HO                      Me
```

derived from the experimental data was undertaken. Again, an energy optimisation using GAUSSIAN86 with a 6-311G** basis set was performed, which showed that the B ... O interaction has indeed a definite covalent character with a "shared" electron count of about 0.68, and a sizeable charge transfer of about 0.4 el from the methanol to the boronic acid.

Conformational Analysis of Peptides Using Molecular Dynamics

G. Barnickel
E. Merck
Frankfurter Str. 250, D-6100 Darmstadt

Peptides, especially peptide hormones and neurotransmitters, play an important role in medicinal chemistry. The study of peptide conformations enables the medicinal chemist to understand the structure-function relationship in order to design compounds which mimic the peptide (peptide mimetics).

In general, however, peptides show a remarkable degree of conformational flexibility which results in numerous possible conformations. For a computational description the multi-minima problem is a serious obstacle.
A complete conformational analysis becomes impossible even for rather small oligopeptides. Therefore different strategies have been developed [1]. Applying these methods a list of energically favoured conformations is obtained; but it remains uncertain whether the global mimimum is found with the approach chosen. Additionally the environment of the peptide (solution, membrane, receptor or crystal) may have a pronounced influence on the relative energy of the calculated conformers. Moreover, the computational efforts for such studies are very time-consuming. For all these reasons a complete conformational analysis has played only a minor role in designing peptide mimetics in industrial research.

An alternative strategy emerges when spectroscopic NMR data become available, which can be used as constraints in molecular modelling. In such cases molecular dynamics calculations (MD) are an appropriate method to find conformations compatible with the geometric data obtained from experiment.

Our research interests are concerned with the tachykinins [2], which seem to play a role as neurotransmitters of pain. They are a group of neuropeptides with similar C-terminal amino acid sequences. Physalaemin is a member of the tachykinin family with the following amino acid sequence:

Glp-Ala-Asp-Pro-Asn-Lys-Phe-Tyr-Gly-Leu-Met-NH2.

A NMR study in DMSO has been undertaken [3]. It was found that at pH 7.0 a β-turn in the region Asp-Lys is stabilized by a salt bridge between the side chains of Asp and Lys. Using the geometrical data of the NMR work a conformational study using molecular dynamics was performed.

In our calculations three different conformations were selected as starting structures. An α-helix and an extended chain both spanning the full peptide as well as a β-turn between Asp and Lys with an extended chain at the C-terminus were chosen for a restraint MD run for 20 picoseconds at 300K. After each picosecond a snapshot was stored followed by an energy minimization. The conformations obtained are depicted in Figure 1. The α-helix does not remain stable during the simulation. The peptide partly unfoldes C- and N-terminally yielding a set of dissimilar conformations. Also the β-strand conformation was altered leading to more or less extended structures. Only in the case of the β-turn start conformation does the overall shape of the peptide backbone remain stable.

Considering only those conformations that fulfil the constraints extracted from the NMR measurements a set of quite different conformations was obtained. Some representative examples are depicted in Figure 2. Although the salt bridge represents a long range interaction this constraint taken together with the distances calculated from the NOE effects allow a variety of conformations. It is concluded that physalaemin in DMSO is rather flexible and the salt bridge found can be accomodated by different conformations including a β-turn conformation.

In order to test whether the salt bridge can be substituted by a convalent bond two cyclic analogues were synthesized:

BOC-cyclo-(Asp-Pro-Asn-Lys)-OMe 1
BOC-cyclo-(Glu-Pro-Asn-Lys)-OMe 2

A 2D NMR study in DMSO was undertaken [4]. The geometrical parameters determined were used for the MD calculations. For both peptides a 100 picosecond MD run with and without constraints was carried out. Along the trajectory after each picosecond a conformation was stored for a subsequent mimization without constraints.

Comparing the energies after minimization it is interesting to note that the conformational energy obtained from the MD calculation without constraints (see Figure 3) are lower than those reached by the constraint procedure. It is concluded that the conformations favoured in vacuum are distinct from those found in DMSO. This finding is of great significance for all kinds of conformational analysis of peptides. A simulation in vacuum may lead to conformations irrelevant to the solution conformation of the peptide under consideration. Therefore a realistic conformational analysis on the basis of molecular dynamics calculations needs the input of experimental data.

The MD calculations were carried out using the DISCOVER program from BIOSYM running on a CONVEX C1. For analysis of data and graphical representation the CPECM program [5] was utilized.

1   Howard AE, Kollman P (1988) J Med Chem 91:1669
2   Hoelzemann G (1989) Kontakte 2:3
3   Hoelzemann G, Pachler KGR (1989) Int J Peptide Protein Res 34:139
4   Hoelzemann G, Pachler KGR, Eberhart B, Hoelzel H, Kraft M, Barnickel G submitted
5   Barnickel G, Dissertation, Freie Universitaet Berlin 1983

Figure 1:

Conformations of physalaemin obtained after 20 picosecond MD with subsequent minimization after each picosecond. All atoms forming the peptide backbone starting from N:Glp to N:NH2 are shown.

a) The α-helical backbone conformation initially used did change during the calculation.

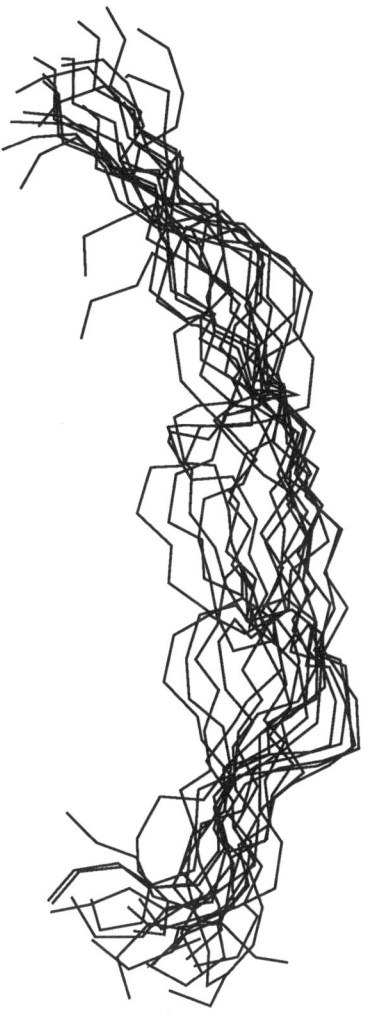

Figure 1:

b) The β-strand conformation changes to more or less extended structures.

Figure 1:

c) In case of the β-turn the overall backbone folding remains stable.

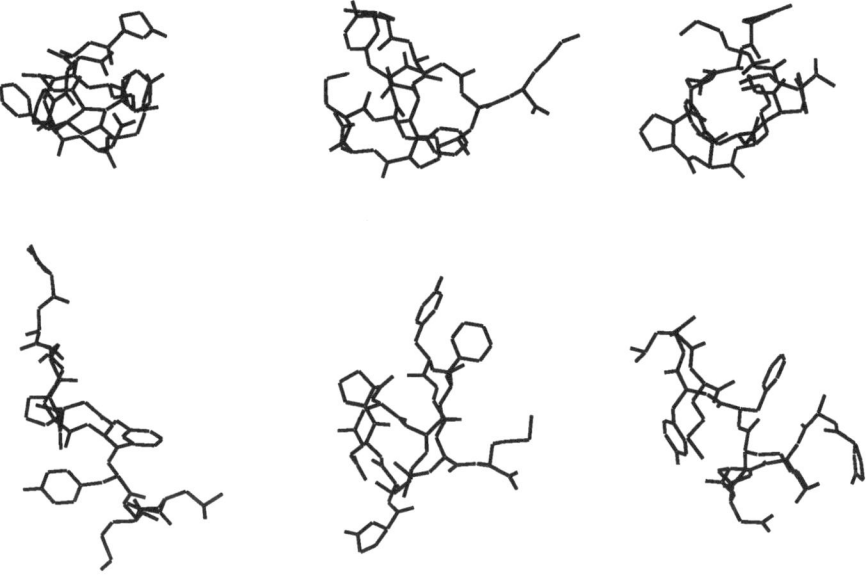

Figure 2:

Representative conformations compatible with NMR data obtained for physalaemin in DMSO solution. In all cases the salt bridge between the charged side chains of Asp and Lys is conserved. The geometrical parameters determined from NMR are consistent with a number of dissimilar conformations.

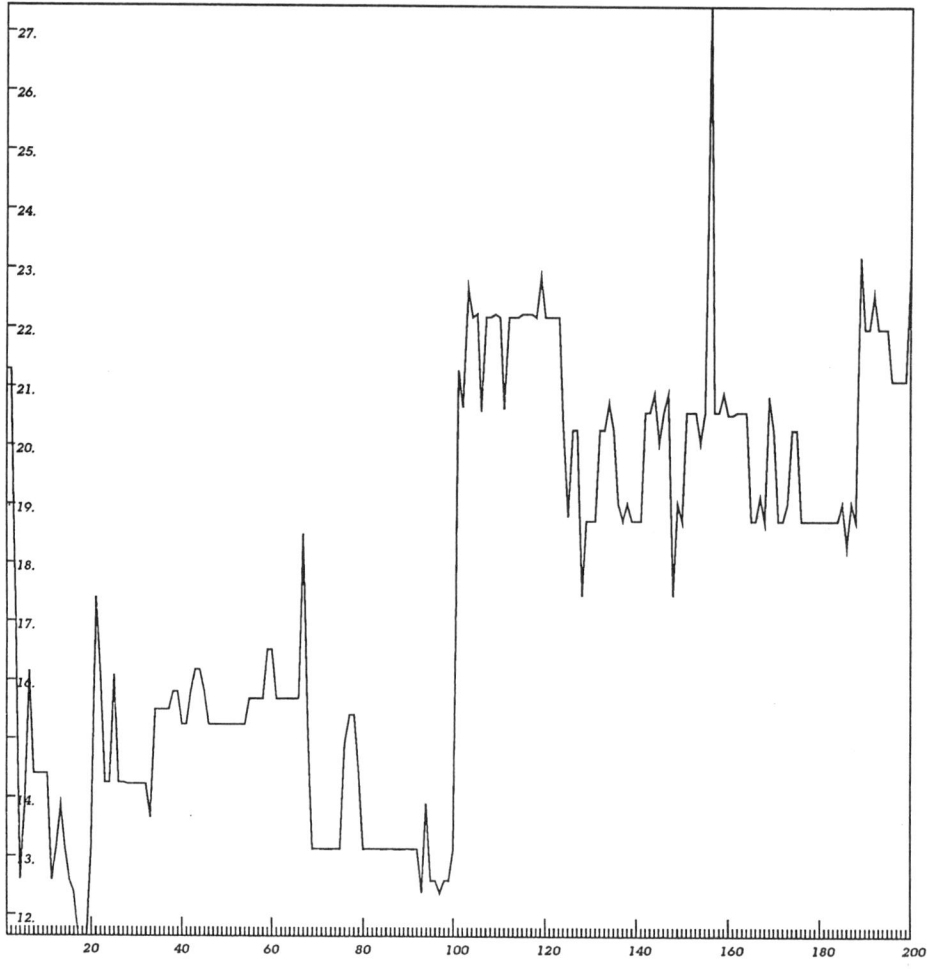

Figure 3:

Energy profile of conformations for Boc-cyclo(Asp-Pro-Asn-Lys)-OMe. Conformations 1 to 100 were obtained from a MD run for 100 picosecond without NMR constraints, 101 to 200 from the constraint run. After each picosecond a conformation was stored and a subsequent minimization was carried out. The procedure without constraints yields conformations with considerably low energies compared to the constraint case.

The Use of Supercomputers in Medicinal Chemistry Examples from Peptide and Protein Projects

H. Köppen
Boehringer Ingelheim KG, Med. Chem. Dept., 6507 Ingelheim

Abstract: Three examples of the use of molecular modeling techniques in drug research are discussed. A model of the threedimensional structure of Neuropeptide Y (NPY) was built up and used as a template for the successful design of a smaller peptide of very similar biological action. In the case of the very flexible ANF (atrial natriuretic factor) a conformational analysis using molecular dynamics at elevated temperatures gave essential hints on the bioactive conformation. In the third example (Phospholipase A_2) some aspects of the modeling of protein structures and the design of active site inhibitors are briefly discussed. Modeling studies in all these examples could be performed only because of the availability of supercomputers.

There is a large number of various diseases which threaten man's life or which impair its quality. Many of them can be cured today by modern drug therapy but there are still therapeutic challenges for which new drugs are needed, e.g. AIDS, malignant tumors or Alzheimer's disease, to name only a few. There are even more diseases where available drugs can treat the symptoms only or where unwanted side effects are observed.

The search for a new drug was and still is a difficult, time consuming and expensive process where luck plays a major role. However, during the past two decades new theoretical approaches were developed aiming at more rational ways of drug research. The scientist's creativity is still one of the most important prerequisites for finding new and safe drugs but statistical analysis (QSAR) and theoretical chemistry (molecular modeling) techniques are available today to assist the scientist in developing and testing hypotheses with respect to the intermolecular interaction of a drug with its target, the receptor or the enzyme.

The computer assisted analysis of the available chemical and biological data may result in a first crude hypothesis which structural elements of the drug are essential for binding and/or specificity (these structural elements form the so-called "pharmacophore"). In most cases the first experimental basis will not be sufficient to un-

U. Harms (Ed.)
Supercomputer and Chemistry
© Springer-Verlag Berlin Heidelberg 1990

ambiguously identify the pharmacophore. Hence new compounds with structural variations will be synthesized and their biological activity will be determined. Molecular modeling techniques allow to calculate structure dependent properties as e.g. energetically stable conformations, molecular volumes, charge distributions or molecular electrostatic potentials. They do not allow to calculate biological activities in contrast to a very common misconception. However, they open a way to design compounds with defined molecular properties. The combined information about these molecular properties and the biological action allows to test and to refine the "pharmacophore" hypothesis which in turn leads to the design of new compounds. In this iterative way hypotheses about structure activity relationships can be refined and used to improve the biological profile of a drug on a rationale basis.

Molecular modeling is basically the application of theoretical chemistry techniques. It is therefore dependent on approximations which must be kept in mind in order to avoid misleading interpretations of the calculated data. The mere use of a fast computer or a high performance graphics display cannot guarantee the quality of the produced or displayed data if the above mentioned rule is neglected. Nevertheless there can be no doubt that the dramatic increase in computer hardware performance during the last years was the major reason for the rapidly increasing number of applications of molecular modeling to problems of realistic size. The availability of fast computers indeed allowed to test and to refine more realistic (and more computational intensive) models which in turn led to a progress in theoretical approaches.

A second important aspect is the increasing amount of experimental data about molecules of biological interest. The most remarkable feature of that development is the progressive information about the threedimensional structure of proteins which are of fundamental importance for all biological processes. To grow a protein crystal is still more an art than a science but computer technology helps to get the atomic coordinates as soon as suitable crystals are available [1]. Protein structures in solution can be determined by the combined use of modern NMR techniques and constrained molecular dynamics calculations [2,3,4]. Gentechnology helps to get the respective protein at hand but also offers a way to study modified protein structures via site directed mutagenesis (see e.g. [5]). Deeper insight into enzymatic catalysis is one important goal of these experiments. But there is another one: The still unsolved problem of the basic rules of protein folding is one of the great scientific challenges today. The rapidly growing amount of data from site directed mutagenesis experiments may eventually provide us with the key informations for an understanding of these rules.

There are recent reviews on modern drug research [6] as well as molecular modeling techniques [7] where more details can be found. In the following three examples of molecular modeling applications in drug research will be discussed. They are taken from peptide and protein projects, respectively, at Boehringer Ingelheim.

The research efforts in these projects are not directed towards peptides used as drugs. In general peptides cannot be orally applied which is the preferred route of administration. Enzymatic degradation in the gastrointestinal tract is faster than absorption in most cases. However, peptides or proteins may serve as a kind of lead structure for rational drug design, either as a target for the small drug molecule or as a model for the design of a small peptidomimetic compound or a peptide inhibitor. In the first case the knowledge of the threedimensional structure of the active site where the drug should bind is helpful for designing enzyme inhibitors. In the second case the small, nonpeptidic compound should mimic the bioactive conformation of the peptide (e.g. a hormone) at the receptor in order to act either as an agonist which triggers the biological response or (by suitable structural modifications) as an antagonist which prevents the docking of the peptidic hormone at the receptor site.

The first example describes the approach taken to design Neuropeptide Y (NPY) antagonists. The second example is dealing with ANF, an endogenous peptide with vasorelaxant properties. The third example demonstrates some aspects of the design of Phospholipase A_2 inhibitors.

NPY

Neuropeptide Y (NPY) is a 36-amino acid peptide amide. It is a member of the family of pancreatic polypeptides [8] which are characterized by high sequence homology. NPY is an important neurotransmitter and one of the strongest endogenous vasoconstrictors. In addition it potentiates the pressor effects of other transmitters and hormones and may be involved in the regulation of cardiovascular functions. It is expected that antagonists which interfere with the receptor binding of NPY may offer an attractive therapeutic route in the treatment of hypertension.

There are no small molecule inhibitors of NPY known so far. For this reason the only "lead" structure in the rational design of NPY inhibitors is the structure of the natural agonist, NPY, itself. The aim is to determine its bioactive conformation in order to derive the spatial arrangement of those functional groups which dominate the intermolecular interaction with the receptor. The next step will be to design a small molecule with a similar orientation of functional groups and to modify the molecular structure of this nonpeptidic molecule in a systematic way in order to suppress the biological action at the receptor without loss in binding affinity. Such a molecule should be a nonpeptidic NPY antagonist.

This is an outline of the basic idea but there is no doubt that it is an ambitious approach. It cannot be expected to provide a straightforward route to success. The determination of the bioactive conformation of a peptide is by no means a routine procedure. Theoretical methods typically neglect the interaction with receptors because of the lack of information about their atomic structure. No direct experimental access to the receptor bound conformation of a peptide has been reported so far. On the other hand it is not guaranteed that alternative ways, e.g. a random screening, will lead to the desired compound in shorter time at lower costs. Strategies in drug research are strategies to optimize the chance for success. No single measure can be expected to be the golden panacea and several promising approaches must be taken in parallel.

Despite those difficulties there are structural features of the NPY molecule which make the rational approach more feasible. The x-ray structure of APP (avian pancreatic polypeptide), another member of the pancreatic polypeptide family, has been determined at high precision [9]. It adopts a folded conformation in crystalline state with strong hydrophobic interactions between an N-terminal polyproline helix and a C-terminal α-helix. It is generally assumed that NPY and APP share common structural features [10,11] because of their high degree of homology. CD measurements on NPY and several

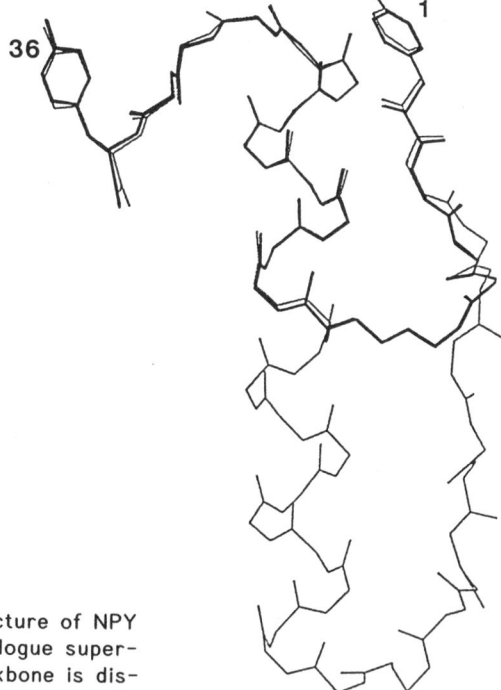

Fig. 1: The hypothetical 3D structure of NPY with the designed small NPY analogue superimposed (bold line). Only the backbone is displayed except residues 1 and 36, respectively.

C-terminal segments demonstrated the α-helical conformation in H_2O/CF_3CH_2OH mixtures [12]. This finding supports the hypothesis that the conformational features found in the APP crystal may be typical of the pancreatic polypeptides even in solution.

Helices and strong hydrophobic interactions provide indeed most important contributions to the overall stability of peptides and proteins. With respect to NPY one might speculate that these rather rigid segments are necessary to hold the bioactive residues in optimum position for the interaction with the receptor. NPY binding studies with mutated C-terminal segments [12] revealed that some of the C-terminal amino acids are indispensable for receptor binding. The isolated N-terminus does not bind to the NPY receptor at all. However, the essential C-terminal amino acids, separated from the rest of the peptide, do not bind either. 10 residues (NPY Ac-27-36) are needed to get at least about 0.1 % of the receptor binding constant of the whole peptide.

The hypothetical 3D structure of NPY (fig. 1), deduced from the x-ray structure of APP by exchange of amino acid side chains and energy relaxation using DISCOVER [13], proved to be very stimulating to the scientists involved in that project. If neither the N-terminus nor the C-terminus alone bind to the receptor with high affinity, they concluded, it might be that both are needed in close spatial vicinity. To test this new hypothesis they designed an NPY analogue consisting of segments 1-4 and 25-36 of NPY linked by a flexible spacer (ε-aminocaproic acid). Modeling techniques were used again for finding the optimum length of the spacer and for answering the question if the new, not yet synthesized compound could fold into the desired shape at all (fig. 1, [14]). The receptor binding test of the drastically shortened NPY analogue clearly supported the hypothesis: Compared with NPY the binding constant of the shortened analogue was only reduced by a factor of 3 [15]. Moreover CD data demonstrated a higher degree of α-helical conformation compared with e.g. Ac-25-36. One might speculate that the receptor binding is somehow correlated with the stability of the C-terminal α-helix.

Molecular dynamics calculations using DISCOVER [13] were intensively used for this project. A 30 picosecond simulation of NPY (time step 1 femtosecond, no cross terms, no morse function, no cutoff, vacuum conditions) takes about 6 CPU days on a VAX 8530. It takes about 1.5 hours CPU time of a CRAY-2 (University of Stuttgart) which was the preferred machine for obvious reasons. This comparison makes clear why fast computers are essential for peptide and protein modeling.

ANF

The atrial natriuretic factor ANF is an endogenous 28-amino acid peptide hormone with potent natriuretic, diuretic and vasorelaxant properties [16]. It is secreted from atrial heart tissue. There are hints that it plays a key role in the regulation of blood pressure. A nonpeptidic compound with the biological profile of ANF may offer a new and very promising therapeutic approach in the treatment of hypertension.

Designing a nonpeptidic antagonist requires some knowledge about the bioactive conformation of the respective peptide as discussed above. This is even more important in the design of a peptidomimetic agonist since both binding and triggering of the biological effect are necessary.

ANF is a medium sized peptide with a disulfide bridge (fig. 2). Aminoacid replacement studies demonstrated that some of the residues within the large loop are essential for biological activity [17]. However, in contrast to NPY there is no threedimensional structure of any homologous peptide known. Moreover NMR studies on ANF under a variety of conditions [18,19,20] failed to detect any preferred solution conformation due to the high flexibility of this peptide. Therefore there are no direct experimental data about the bioactive conformation of ANF which could serve as a starting point for rational drug design.

In an attempt to overcome that problem Mackay et al. [21] generated families of low-energy conformations of ANF using high-temperature dynamics calculations under vacuum conditions. This approach will be discussed in more detail now.

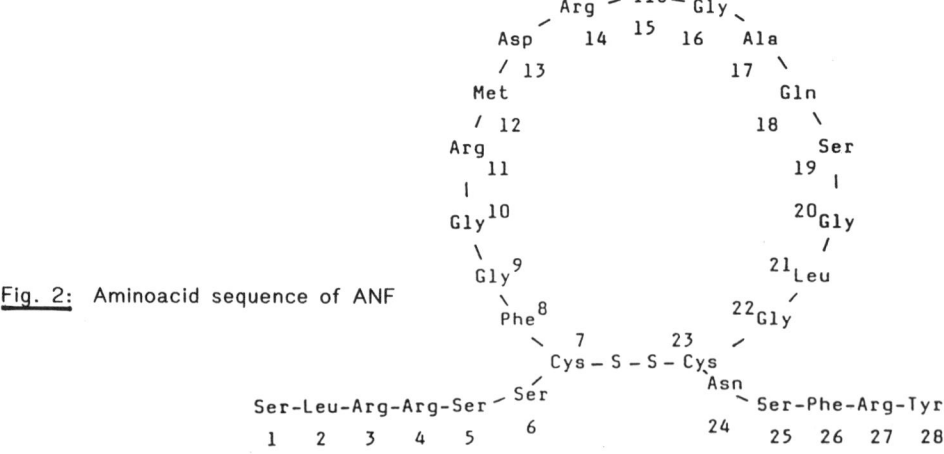

Fig. 2: Aminoacid sequence of ANF

In the case of small molecules the variation of torsion angles about flexible bonds can be used to systematically search the accessible conformations of the respective molecule. Their energy can be determined by conventional energy minimization techniques. The number of conformations which have to be analyzed goes up with m^n where m is the number of torsion angle increments and n the number of flexible bonds. Because of the rapidly increasing computational requirements this approach is limited to 7-10 flexible bonds in practice. Even a small peptide has more flexible bonds and the systematic conformational analysis is therefore not feasible for peptides or proteins, even if future possible improvements of computer performance are taken into account.

During the last years molecular dynamics (MD) calculations became the favourite computational tool for studying the dynamics of proteins and peptides [22]. Due to the simulated thermal motion of the atoms conformational changes may be observed and, by taking snapshots from time to time, selected conformations may be sampled and subsequently analyzed. However, in contrast to the systematic search the results of MD calculations depend on the starting conformation, the simulation time and the temperature. The ability of MD calculations to search the conformational space is frequently overestimated. This can be demonstrated by using a small test molecule where both systematic search as well as MD calculations can be performed.

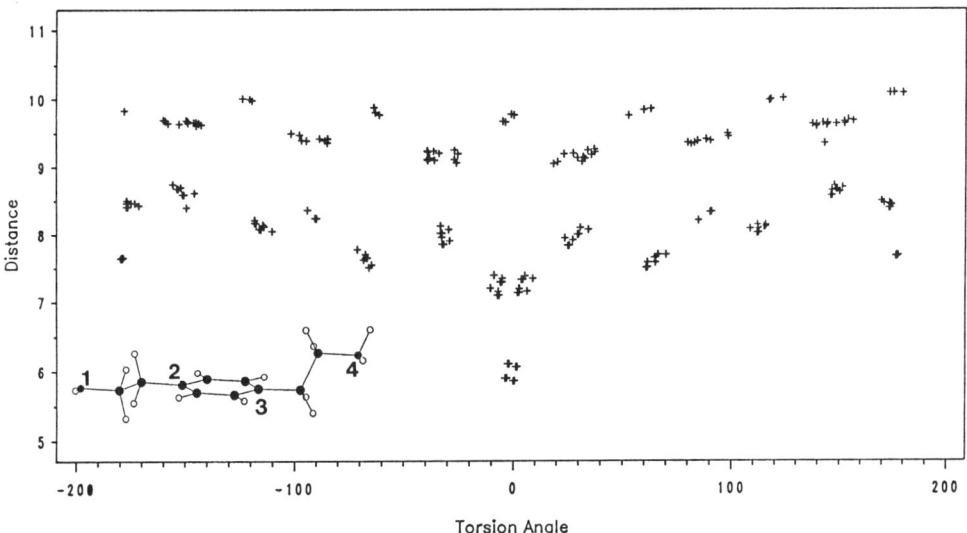

Fig. 3: Result of a systematic conformational search of the test molecule displayed in the insert. The distance (in A) between the *oxygen 1* and *nitrogen 4* is plottet versus the torsion angle *oxygen 1 - carbon 2 - carbon 3 - nitrogen 4*. Only conformations with a maximum energy of 4 kcal/mol above the lowest energy conformation found are plottet.

A simple parasubstituted benzene ring with with 6 flexible bonds was chosen for test purposes (2-[4-(2-aminoethyl)phenyl]ethanol). The modeling program CHEM-X [23] was used for input structure generation, energy minimization and systematic search (atomic charges were set to 0). In order to display the conformational variability of the molecule the distance between the oxygen and the nitrogen, located at opposite ends of the molecule, was plottet versus the pseudo torsional angle oxygen 1 - ring carbon 2 - ring carbon 3 - nitrogen 4 (see fig. 3). Only conformations with an energy of less than 4 kcal/mol above the lowest energy minimum found are plotted. The calculated torsion angles range from -180 to 180 degrees and the distances are between about 6 and 10 Angstroms. The plotted conformations are taken from the systematic search without further energy minimization.

To compare the systematic search with molecular dynamics the input structure of the systematic search was used for a DISCOVER MD run [13]. Dynamics calculations were performed over 100 picoseconds at 300 and 1000 K, respectively, under vacuum conditions without cutoff radius and a time step of 1 femtosecond was used. Charges were set to 0. Snapshots of the generated conformations were taken every single picosecond and the geometric parameters described above were determined for each of these conformations.

The results for the 300 K and the 1000 K MD run, respectively, are displayed in figure 4 (a and b, respectively). In contrast to the systematic search the results of MD calculations depend on the starting conformation which is therefore also displayed in figure 4.

Obviously the dynamics over 100 picoseconds at 300 K (fig. 4 a) gave by far no exhaustive search of the conformational space. Instead there is a strong tendency to stay close to the extended starting conformation with some torsion angle changes which however cannot significantly alter the distance between oxygen and nitrogen. Basically the same is true for the high-temperature dynamics (fig. 4 b) where only the torsion angle range is more densely populated in comparison with the 300 K dynamics. The observed maximum distances between oxygen and nitrogen are slightly larger compared with the systematic search which is most probably due to the fact that dynamics allows for bond angle and bond length changes, respectively, in contrast to the search where these geometric parameters are kept constant.

To study the effect of the starting conformation on the dynamics results a second set of MD calculations at 300 K and 1000 K, respectively, was performed starting from the energy minimum conformation found in the systematic search (fig. 5 a and b). All other parameters of the dynamics protocol were kept constant. Again the covered range of distances is very limited. At 300 K also the torsion angles focus on a narrow range which interestingly differs from the starting value around 0 degrees. Obviously there is

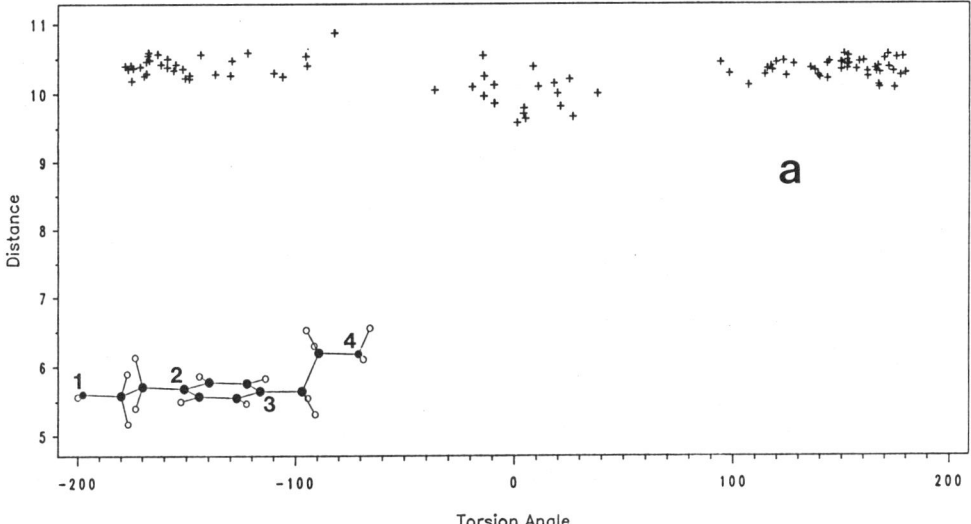

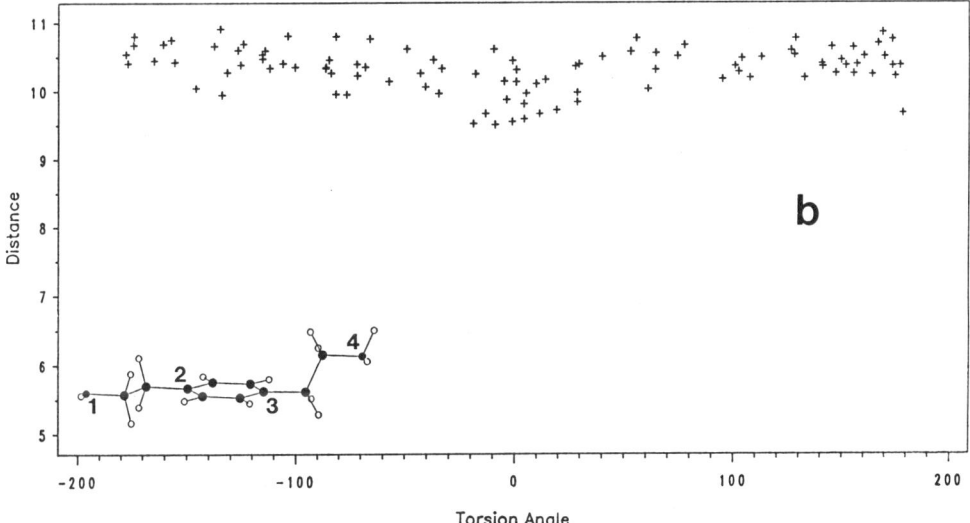

Fig. 4: Distance between *oxygen 1* and *nitrogen 4* (in Å) of the test molecule displayed in the insert versus torsion angle *oxygen 1 - carbon 2 - carbon 3 - nitrogen 4*. The conformations were generated as snapshots taken every picosecond during a 100 picosecond molecular dynamics calculation. Starting conformation as displayed.

a) temperature = 300 K   b) temperature = 1000 K

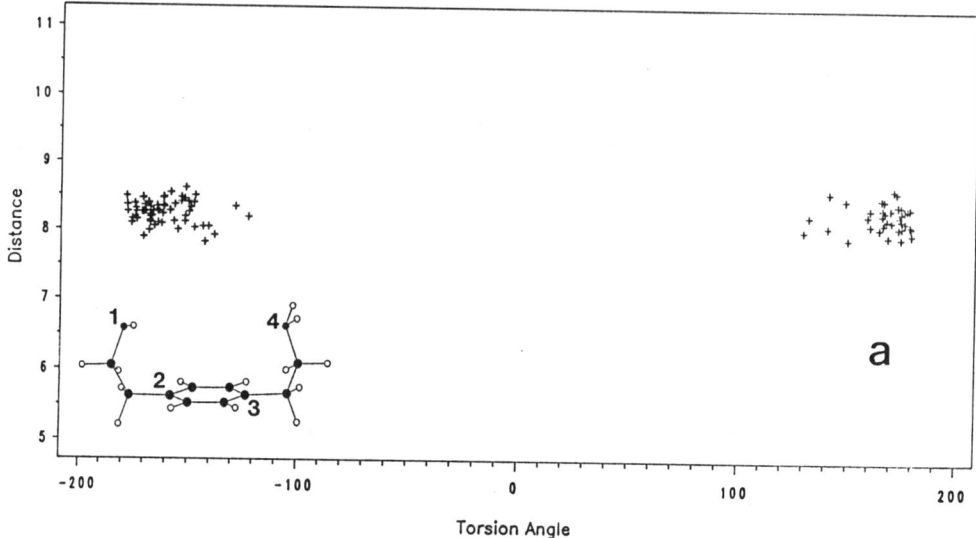

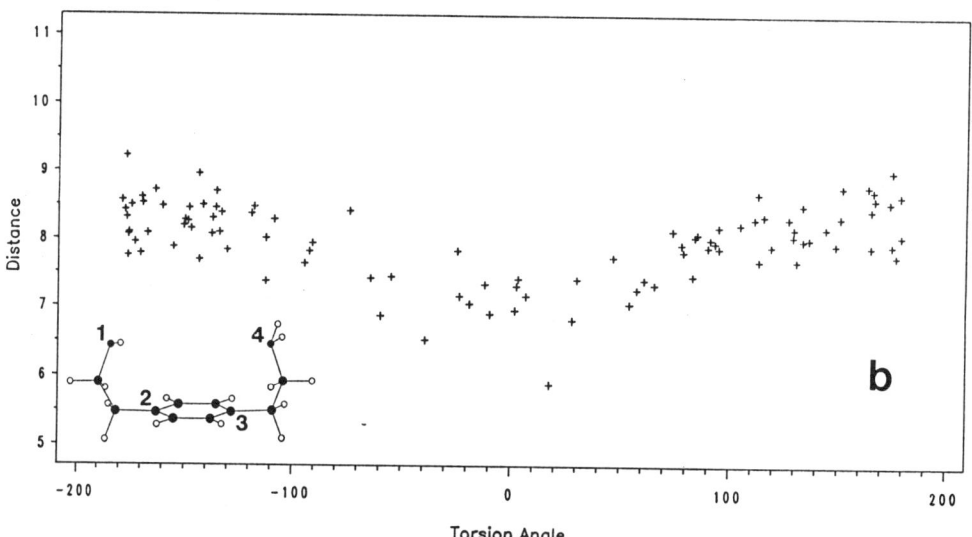

Fig. 5: Distance between *oxygen 1* and *nitrogen 4* (in Å) of the test molecule displayed in the insert versus torsion angle *oxygen 1 - carbon 2 - carbon 3 - nitrogen 4*. The conformations were generated as snapshots taken every picosecond during a 100 picosecond molecular dynamics calculation. Starting conformation as displayed.

a) temperature = 300 K   b) temperature = 1000 K

a major torsion angle change in the order of 180 degrees taking place at the beginning of the dynamics. At higher temperatures a broader range of different conformations is generated but only the combined results of both 1000 K - MD runs give an impression of the range of energetically accessible conformations which is nearly as complete as the systematic search.

The results of the various calculations performed on this simple test molecule clearly demonstrate that molecular dynamics is only of limited value as a tool for searching the conformational space. This is obviously even more true for larger molecules like peptides and proteins. In any case MD results must be interpreted with great care and caution. High-temperature conditions are generally advisable and one must be aware that 100 picoseconds are only a short time range compared with typical relaxation times of larger conformational changes [22].

ANF is a flexible molecule and experimental data about preferred conformations are not available up to now as mentioned above. A systematic conformational search is not feasible because of the large number of flexible bonds. In an attempt to get at least a feeling for the range of accessible conformations Mackay et al. [21] performed 800 K - MD calculations on ANF over 100 picoseconds under vacuum conditions starting from an extended backbone conformation with a beta turn at GLY 16. In order to increase the sampled conformational space the authors randomized the atomic velocities every 5 - 10 picoseconds. Conformations were sampled every picosecond and minimized. For the purpose of comparison the backbone of each structure was optimally superimposed onto the backbones of the other 99 minimized structures. The root mean square deviations (rms) between the superimposed atoms were taken as a quantitative measure of the 3D similarity of these 100 conformations. 11 groups of similar conformations could be formed according to their low rms values.

High-temperature dynamics has a certain bias towards the generation of high-energy conformations. In order to relax structures which might have been trapped in high-energy minima annealing MD calculations at 300 K were performed for representative conformations of each of the 11 structural families.

In the light of the small molecule test case discussed above there can be no doubt that a 100 picosecond MD calculation on a 28-aminoacid peptide cannot exhaustively sample the conformational space even if high temperatures and periodic randomization are used. From this point of view free MD calculations on flexible peptides are often criticized as yielding only irrelevant results. Moreover the neglect of solvent interaction is an artificial condition with significant influence on the generated conformations. Among other things there is a tendency towards compact structures with the maximum number of intramolecular contacts. This effect is more pronounced in low temperature calculations while higher temperatures at least partly compensate for these interactions. However, the mere inclusion of water molecules into the MD calculation does not lead to a higher reliability of the generated conformations in terms of bioactive conformations as the interactions between a peptide and its receptor are highly specific and different from interactions with bulk solvent even if single water molecules may be involved at the receptor site.

It must be stressed that there is no theoretical procedure available for the determination of bioactive conformations just by searching the conformational space. If a systematic search is not feasible one can only try to sample as many different conformations as possible, hoping to find the bioactive conformation among them. Supplementary experimental data are needed to differentiate between the many inactive and the single active conformation. NPY is a good example where the search for the active conformation is facilitated both by a rather rigid peptide and experimental data about its 3D structure.

Being aware of these limitations the scientists involved in that project analyzed the 11 conformations determined by Mackay et al. [21] by a careful comparison with the available structure activity data on modified ANF structures [26]. Additional chemically constrained peptides with a limited flexibility were synthesized and their biological activity was determined. At the end all experimental data were in agreement with the hypothesis that there is a conformation among the 11 sampled structures which is at least close to the bioactive one.

This remarkable result makes clear that there is a chance to find the bioactive conformation by the procedere used by Mackay and coworkers. However it would be wrong to assume that the bioactive conformation can always be trapped by performing a 100 picosecond dynamics at 800 K with periodic randomization. Some luck is needed. And even more important: enthusiastic chemists are needed who are willing to do all that time consuming, tedious synthetic work which is needed to test a hypothesis.

PLA_2

The phospholipases are a group of widespread enzymes which catalyze the hydrolysis of ester bonds in phospholipids. The phospholipases A_2 cleave the ester bond at the central carbon atom of the glycerol backbone of phospholipids. They are playing a key role in inflammation processes. For that reason inhibitors of phospholipases A_2 could be very useful drugs in a variety of diseases like asthma or psoriasis to name only a few.

There are intracellular and extracellular phospholipases. However, only some x-ray structures of the more abundant extracellular phospholipases are known. A remarkable feature of all known phospholipases A_2 is the high degree of homology within the active site. The rational design of active site inhibitors may therefore start from phospholipases with known x-ray structure. In a procedure called "homology building" the protein with known x-ray structure is converted into the desired protein by replacement of amino acid sidechains and by insertion or deletion of residues if necessary. The resulting structure is typically highly strained and must be slowly relaxed in order to avoid artefacts. There are only relatively few threedimensional structural motifs found in natural proteins. However, the huge number of possible energy minima imposes limitations on a model built up by homology building which must carefully be taken into account.

If a model of the threedimensional structure of the protein is available inhibitors may be designed and docked into the active site. Both design and docking are no routine procedures. In particular the design process requires the creativity of scientists. Informations about those points within the active site which favourably bind atoms or functional groups commonly found in drug molecules (e.g. NH_2 or CO) proved to be very helpful. Peter Goodford's program GRID [24] provides these informations. A probe atom is moved around the protein and the interaction energies are determined at points in space forming a threedimensional grid. In this way the position of water molecules found for instance in the x-ray structure of the porcine pancreas phospholipase A_2 [25] can correctly be predicted.

After docking the potential inhibitor into the active site the system must energetically be relaxed by allowing the protein and the inhibitor atoms, respectively, to adjust their positions in order to find a new energy minimum.

There are various possible orientations of a molecule in the active site of an enzyme. One has to keep in mind that in most cases docking cannot provide a unique answer how the most powerful inhibitor must look like. Instead it stimulates the creativity of the scientists and gives hints which structures probably will not fit the active site and should therefore be avoided. As already described above the feedback from experiment, in particular from biological tests or (if feasible) from x-ray analysis, is urgently needed in order to refine the model.

CONCLUSIONS

Three examples of modeling applications in drug research have been discussed. A common feature of all these research projects is the use of supercomputers like CRAY-2, STAR 100 or CONVEX C-220. While some modeling of smaller peptides can in principle also be done on slower machines fast computers are an indispensable tool for the handling of larger molecules (e.g. the porcine pancreas phospholipase A_2 consists of about 1800 atoms and in addition about 500 water molecules are explicitly included in all calculations). However that describes only a part of the benefit from the use of fast computers. All modeling procedures are based on approximations. The quality of results can only be assessed if several calculations with a systematic variation of parameters can be performed. In this way fast computers can be very helpful even for less computational intensive procedures and be used by the scientist like a new kind of "experimental" device.

ACKNOWLEDGEMENTS

Dr. Ralph Anderskewitz, Dr. Horst Dollinger and Dr. Gerd Schnorrenberg are acknowledged for many stimulating discussions. Dr. Klaus Angermund is thanked for his assistance in using DISCOVER on a STAR-100 computer.

REFERENCES

1 Gros P, Fujinaga M, Dijkstra BW, Kalk KH, Hol WGJ (1989) In: van Gunsteren WF, Weiner PK (eds) Computer simulations of biomolecular systems. Escom, Leiden
2 Wuethrich K, (1989) Science 243:45
3 Kessler H, Bats JW, Wagner K, Will M (1989) Biopolymers 28:385
4 Clore GM, Gronenborn AM (1989) Crit Rev Biochem Molec Biol 24:479
5 Rossi J, Zoller M (1987) In: Oxender DL, Fox CF (eds) Protein Engineering. Liss, New York
6 Martin YC, Kutter E, Austel V (eds) (1989) Modern Drug Research. Marcel Dekker, New York, Basel
7 Cohen NC, Blaney JM, Humblet C, Gund P, Barry DC (1990) J Med Chem 33:883
8 Glover ID, Barlow DJ, Pitts JE, Wood SP, Tickle IJ, Blundell TL, Tatemoto K, Kimmel JE(1984) Eur J Biochem 142:379
9 Glover I, Haneef I, Pitts J, Wood SP, Moss D, Tickle I, Blundell T (1983) Biopolymers 22:293
10 Glover I, Blundell T (1983) Top Mol Pharmacol 2:123
11 Allen J, Novotny J, Martin J, Heinrich G (1987) Proc Nat Acad Sci USA 84:2532
12 Beck A, Jung G, Schnorrenberg G, Gaida W, Lang R (1989) In: Jung G, Bayer E (eds) Peptides 1988. Walter de Gruyter, Berlin, New York
13 BIOSYM Technologies, San Diego, USA
14 Beck A, Jung G, Gaida W, Köppen H, Schnorrenberg G, Lang R, submitted to publication
15 Beck A, Jung G, Gaida W, Köppen H, Lang R, Schnorrenberg G (1989) FEBS Lett 244:119
16 deBold AJ, Borenstein HB, Veress AJ, Sonnenberg AT (1981) Life Sci 28:89
17 Nutt RF, Veber DF (1987) Endocrinol Metab Clinics North America 16:19
18 Fesik SW, Holleman WH, Perun TJ (1985) Biochem Biophys Res Comm 131:517
19 Theriault Y, Boulanger Y, Weber PL, Reid BR (1987) Biopolymers 26:1075
20 Gampe RT, Connolly PJ, Rockway TW, Fesik SW (1988) Biopolymers 27:313
21 Burt SK, Mackay D, Hagler AT (1989) In: Perun TJ, Probst CL (eds), Computer-Aided Drug Design. Marcel Dekker, New York, Basel
22 McCammon JA, Harvey SC (1987) Dynamics of proteins and nucleic acids. Cambridge University Press, Cambridge (Great Britain)

23 Chemical Design Ltd, Oxford, Great Britain
24 Goodford PJ (1985) J Med Chem 28:849
25 Dijkstra BW, Renetseder R, Kalk KH, Hol WGJ, Drenth J (1983) J Mol Biol 168:163
26 Schnorrenberg G, unpublished results

Local Density Functional Calculations on Properties of Large Molecules

B. Delley
Paul Scherrer Institut, c/o Laboratories RCA, Badenerstrasse 569, CH-8048 Zürich

Abstract: The local density functional is a well tested way to calculate quantum mechanical properties of atoms, molecules and solids. The Dmol implementation of such calculations is briefly introduced. A short overview on applications is given. Summaries of systematic studies are shown to guide expectations on the performance of the approach. Some applications are shown to illustrate the present capability.

INTRODUCTION and METHOD

Today chemical industry active in research is increasingly interested in methods which rely on mathematical, numerical tools. Since medium and large molecules dominate in applications, it is necessary to make use of drastical approximations in order to make the problem amenable to calculation. Molecular modelling may already be regarded as a classic tool in the study of large molecules. But also semiempirical quantum mechanical methods form already part of the methods used routinely. Such methods are currently improved to permit wider applications. However, one basically wants a theory as complete as possible, not depending on parametrizations of experimental data. Today's ab initio quantum mechanical methods only partially fill the gap, since available computing power admits only punctual studies. This is especially true if correlation effects have to be included in the calculations.

The density functional methodology and in particular the local density functional (LDF), first introduced by Kohn and Sham [1], has a long standing tradition now in solid state physics. But there have also been many applications of the LDF in chemistry. Contrary to the Hartree Fock (HF) methodology both exchange and correlation are modeled at a consistent level by an effective single particle potential. This admittedley approximate procedure avoids the pathology of the Hartree Fock approach. It permits interpretation of theoretical results in terms of orbitals which generate the density. Well designed density functional methods scale in computing time with the cube of the system size for extremely large systems. While calculations for intermediate molecules may fit to a lower power law, the asymtotic behavior is dominated from the cubic term which appears by necessity in a method resulting with orthogonal orbitals.

An early application of the LDF, the MsXα (Ms: multiple scattering, Xα: a particular form of the LDF) has found wide distribution. However, additional approximations to the static potential, which are unrelated to LDF make this method unreliable.

The recently introduced Dmol method is a numerical, variational implementation of the Density functional for molecules. Despite that it has not yet come to age it is already used in industrial research. Its ancestors [2] date back more than a decennia and have helped to establish the usefulness of the approach.

The Dmol method is a high accuracy method for polyatomic molecules and clusters. The inherent high accuracy makes it easy to get valid results and makes the need for expert knowledge for picking an adequate variational and static-potential basis set obsolete. The method derives its merits from four interrelated specialties.

First, the method uses easy to handle numerically defined variational functions, which guarantee exact solutions in limiting cases. For normal molecular calculations the numerical basis sets are exceptionally small for a required quality level. Experience shows that the optimum is nearly reached in all cases with just a standard basis set for that accuracy level. No tailoring of the basis set is needed for each section of the Born Oppenheimer surface. And there is little room for basis set superposition errors.

Second, a three dimensional fully automatic integration scheme is used to calculate the matrix elements of the algebraic eigenvalue problem. This allows in principle the use of any type of basis function consistent with the basic analyticity conditions for molecular orbitals.

Third, a perfectly flexible representation of the static and exchange correlation potential is obtained in a N^2 algorithm. This makes the use of a density fitting basis obsolete.

Finally, the algorithm incorporates analytical derivatives of the total energy function. The extra effort needed to calculate derivatives is smaller than the one for three extra iterations in the LDF procedure and scales like N^3. This fully vectorized algorithm is in the process of beeing incorporated into a geometry optimizer routine. The benefits of such force calculations were not yet available for the examples presented here for the Dmol method.

It is possible to make full use of symmetry by providing Dmol with the transformation matrix for the transformation of the atomic orbitals to the symmetrized basis orbitals. The computer program has a general structure aiding parallelization and vectorization. There exists a version performing at Gigaflop speed on multiprocessor Cray computers. The Dmol code has been described in a publication in press [3].

Basically the numerical approach to solving the LDF equations has drastically reduced the need for expert knowledge on basis sets, fitting expansions etc to obtain an acceptable approximation to the exact LDF solution on a particular point on the energy surface.

OVERVIEW on APPLICATIONS

LDF methods in general have been used in applications on a large variety of compounds. Table 1. gives an overview of applications that have been published or will be published. A variety of computational methods have been used.

Table 1: LDF Applications:

Molecules
  organic and anorganic molecules
  organometallic
  polymers
  charge transfer compounds
  clusters

Solid state
  alloys
  ceramics
  zeolithes
  Point defects in metals, semiconductors and insulators

Surfaces and Interfaces
  clean and with adsorbates

Also a considerable number of properties has been investigated by LDF methods in the past. Certainly the list in Table 2 is not exhaustive and should rather be taken as representative for the possibilities. A molecular LDF code can be used, apart for calculations of molecular properties, also for studiing properties of cluster models for surfaces and bulk. However, bulk and slab perfectly crystlline periodic structures are currently a domain for specialized LDF codes designed for that kind of symmetry.

Table 2: LDF Investigated Properties:

Electron density
  electrostatic moments
  polarizabilities
  spin density
  magnetic properties

Energy surface
  molecular geometries
  adsorption / chemisorption sites
  vibration frequencies
  rotation barriers
  transition states

Excited states
  spectra
  electronic transport in metals

PERFORMANCE of the LDF

It is interesting to consider the avarage quality obtained by LDF methods for organic molecules. Versluis and Ziegler have done a comprehensive study [4] summarized in Table 3. They actually use Slater type orbital sets and the $X\alpha$ form of the functional. The author feels on the basis of systematic considerations and on punctual studies that small numerical basis sets would yield similar but not inferior agreement to experimental data.

Table 3: Bond length errors for organic molecules
with polarized basis set [4]

R	LDF_α	HF
C-C	0.005Å	0.014Å
C-H	0.008	0.008
C-O	0.005	0.015
C-S	0.005	0.010
C-N	0.007	0.015
C-P	0.008	0.015

Clearly the group cited would claim about twice as good average agreement with experimental data for the bond lengths as Hartree Fock methods and (not shown) about the same quality as HF for the angles. Table 4 shows their evaluation for the geometries of transition metal complexes. Comparison is also made with HF and with correlated calculations at the MP2 level. In view of earlier experience with LDF on transition metal compounds the choice of the $X\alpha$ form is not harmless. In the case of the Cr_2 molecule this approximation is known to yield way too large bond distance.

Table 4: Bond length's for Transition Metal complexes
with polarized basis set

molecule	geometrical parameter	LDF	exp.	HF	MP2
CuCl	R(CuCl)	2.09	2.05		
Ni(CO)$_4$	R(NiC)	1.794	1.825	1.921	
	R(CO)	1.141	1.122		
Fe(C$_5$H$_5$)$_2$	R(FeC)	1.65*	1.65	1.88	1.72
HCo(CO)$_4$	R(CoC$_{ax}$)	1.753	1.764		
	R(CoC$_{eq}$)	1.779	1.818		
	R(CO)$_{av}$	1.144	1.141		
	R(CoH)	1.516	1.556		
	$\theta(C_{eq}CoC_{ax})$	97.6	99.7		

quoted from [4], except * Fitzgerald unpublished

Clearly LDF is in rather good agreement with the cristallographic data, while HF based methods underestimate d-bond strength considerably with a concomittant overestimation of bond length. The ferrocene example calculated with Dmol suggests that correlation has to be included to a high degree in wavefunction expansions to match the LDF result for transition metal complexes.

EXAMPLES

Chemisorption Studies

The first case [5] concerns the electronic structure, adsorption geometry and vibrational frequency of single Cu or Ag atoms on a Si(111) surface. This surface is of interest in the synthesis of dimethyldichlorosilane from elemental silicon and methylchloride. This reaction is catalyzed primarily by Cu. Interestingly Ag shows no catalytic activity. Thus, a theoretical comparison of Cu and Ag interacting with a Si surface is expected to provide insights complementary to the ones gained from experiments. Binding energy results show that both Cu and Ag adsorb in the threefold hollow sites with equilibrium heights of 0.74Åand and 1.48Åabove the plane of surface Si atoms. The adsorption energies are found to be 92kcal/mol and 72kcal/mol. Taking a rigid substrate, the calculated frequencies of the perpendicular vibrational modes are 58cm^{-1} and 90cm^{-1}. Lateral diffusion barriers are 12kcal/mol and 8kcal/mol. Calculations for Cu or Ag being moved towards the interior of the cluster, letting the nearest-neighbor Si atoms relax, demonstrate that Cu has a much lower penetration barrier than Ag: 4 vs 53kcal/mol. Therefore, at elevated temperatures, Cu can be expected to penetrate through the silicon surface, whereas Ag would remain above the surface. Adsorbate induced electron density differences indicate that Cu weakens the bonds between surface and subsurface silicon atoms. Ag has a significantly smaller effect. The calculated results suggest that the catalytic activity of Cu and the absence of activity of Ag in the synthesis of methylchlorosilanes is possibly due to the ability of Cu to penetrate into the Si surface forming the intial stages of copper-silicide, whereas Ag stays at the surface and desorbs at higher temperatures.

A second example concerns chemisorption of K and O on a Si(001) surface [6]. It is known that potassium promotes oxidation of the Si surface, it can be desorbed at moderate temperatures. The possibility of forming conducting chains of alkali metal atoms at the Si surface has added much interest to this system. However, some results are still controversial, pointing to fundamental questions regarding the understanding of metal-semiconductor bonding. A main disagreement is whether the K-adsobate forms a conducting chain or a nonconducting ionic chain. In the second case the Si surface would gain charge from the atoms and become metallic. Table 5 shows some of the statements about the K-Si surface system.

Table 5: K and O_2 on the Si(001)2x1 surface

theory & facts			
group	d(K-Si) [Å]	bond type	chemisorption site
Levine 1973			pedestal model CsSi(001)
Ciraci Batra	2.59[Å]	ionic	pedestal
Ishida & al	3.52	0.1 e	pedestal
Tsai Kasowski	3.32	metallic	pedestal
Tochihara & al exp	phys	metallic	same leed as Si(001) 2x1
SEXAFS exp	3.14 ± 0.1		

The Table shows that conceptions about this system disagree wildly, allowing for a wide range of bond length and bonding type. However previous work apparently did not check the chemisorption site. Cluster models with up to 89 atoms have been investigated theoretically to simulate various chemisorption sites. Table 6 shows that alternative chemisorption sites are at least 9kcal/mole less bound the preferred cave site.

Table 6: K and O_2 on the Si(001)2x1 surface:

Results on alternative chemisorption sites for K			
site	energy	d(K-Si)	
cave	0.00eV	3.22Å	*preferred site*
valley bridge	0.40	3.54	
pedestal	0.70	3.21	
bridge	0.75	3.11	
charge transfer 0.53e K → Si			

The charge transfer from K to Si was found to be always near 0.5e, which is rather high for an SCF calculation. The K-Si bond length for the cave site is in agreement with the latest surface enhanced extended X-ray absorption fine structure (SEXAFS) measurement (3.14±0.1Å) Calculations have also been done for O_2 and O_2 with K on this Si surface using appropriate cluster models. O_2 is adsorbed at the bridge site with height 1.90 Å. Co-adsorption of K and O_2 results in an increase of the K height by 5% and an increase of the net binding energy by 0.23 eV/O_2 and increase of charge transfer onto O_2 from .32e to .56e. This clearly implies a weakening of O_2 bond, which is also evidenced by a decrease of the O_2 vibrational stretch frequency by 8%. The coadsorption study thus supports experimental observation of the catalytic promotion of silicon oxidation by alkali metals.

Energy Surface for Light Atoms in a Metal

The investigation of the energy surface for nuclei approaching each other to very close distance in a fusion process is an application of the LDF approach to extreme geometries [7]. The principal question to be discussed is whether loading of Pd with Deuterium can promote a significant spontaneus fusion rate. A cluster model for the solid was used to determine if electronic screening in Pd is sufficient to allow for a greatly enhanced tunneling rate. Vibration function tails could be calculated down to small distance using the LDF energy surface. The pair binding energy in the host has been used to estimate the thermal concentration of D_2 and other potential reactants. Nuclear reaction-crossections at low energies have been used to extract the nuclear reaction constants. Putting these ingredients together gives estimates of the spontaneous reaction rates. It was found that adiabatic electronic screening enhances the tunneling amplitude at small separations by many orders of magnitude. The screeneing effects due to the host weaken the electronic binding leading to lowest energy configurations with large separations. for light nuclei these effects mostly compensate; see Table 7.

Table 7: Cold Nuclear Fusion Results

nuclei	$R_{eq,vac}$	$R_{eq,Pd}$	$\log \Lambda_{vac}$	$\log \Lambda_{Pd}$
dp	0.74 Å	1.70 Å	$-55s^{-1}$	-55
dd	0.74	1.70	-63	-71
dt	0.74	1.70	-68	-77
^{6}Lid	1.60	1.75	-197	-172
^{7}Lip	1.60	1.75	-149	-131

Also the calculated conformational energies tell that the thermal population of short bond-length conformations remain extremely small. On the whole these calculations suggest that the spontaneous occurrence of nuclear fusion in D_2 loaded palladium at room temperature would remain unobservable.

Charge Transfer Salt

The charge transfer salt formed from the electron donor, tetrathiafulvalene (TTF), and the electron acceptor, tetracyanoquinodimethane (TCNQ), has raised high interest since the demonstration of electrical conductivity. The crystal structure is characteristic of the organic conductors of this type with a formation of segregated stacks of donor and acceptor molecules. This apparently unlikely separation of charges must represent the stable ground state despite its electrostatic energy. The conductivity is very anisotropic in the cristalline state it is high along the stacks of like molecules. A study of the detailed intermolecular interactions appears to be essential for understanding stack formation and the origin of electrical conductivity. A theoretical study [8] has yielded a number of interesting results. First it was shown that like molecules do have significant binding:TCNQ-TCNQ 1.05eV, TTF-TTF 0.65eV. And intermolecular distance and slip parameter are very similar to the crystal geometry at the energy minimum for the pair in vacuum. The TTF-TCNQ pair bind only by 0.1eV in the crystal geometry. Binding in charged dimers and trimers of both TTF and TCNQ has been calculated. Most striking is the stability of TCNQ dimers and trimers carrying two and three negative charges already in the absence of counter ions. Connection can be made to electronegativity theory. For the crystal a charge transfer of 0.6 electrons is found at the equilibrium point. The qualitative molecular orbital picture can be described as follows. The stabilities of the stacks are due intermolecular bonding. For long stacks the eigenvalues of the bonding orbitals form bands. The homo energy band for the donor and the lumo band for the acceptor overlap in energy, under which conditions charge transfer further stabilizes the structure. Quite naturally such charge transfer is fractional, the bands remain fractionally filled at equilibrium. This is the characteristic of a "one dimensional metal". For macroscopic samples unoccupied delocalized states occur infinitely closely in energy to the occupied ones. Application of an electric field changes the occupation of states with a direct current resulting. It should be observed, however, that some small coupling in the other directions is necessary, otherwise the conductance would be suppressed by fluctuations.

Electron Density of Organic Molecules

In several studies the electron distribution in space has been the object of study. Three sulfur containing molecules that exhibit unusual S...O contacts have been studied with LDF calculations, extended Hückel and crystallographic methods [9]. Theory has helped to analyze the interactions in the X-S...O-Y relevant part of the molecules. Although good cristallographic density maps are very hard to obtain the joint study has helped understand the bonding interactions. There turns out to be a competition of various σ-type couplings: binding of the oxygen lone pair through the antibonding X-S orbital with stabilization through sulfur d orbitals that counteracts the repulsion between the oxygen lone pair and filled orbitals around sulfur. According to this result the S...O distance depends on the electronegativities of X and Y.

In the case of the tetrafluorotrephthalonitrile [10] and the Boron-nitrilotriacetate [11] the electron density has been studied using an extended basis set including functions suitable for core polarization. Static density maps have been compared to X-ray maps with very high resolution. The charge densities in the bond regions are discussed in terms of peak height and bond polarity. The more electron rich the atoms are the lower are the peaks. Bond peaks tend to be displaced towards the more electronegative atom. In contrast ot chemical expectation B-N appears to be more polar than B-O and C-N more polar than C-O. This is due to the representation of the bonding density as $\rho(observed)$-$\rho(promolecule)$. While for an exact LDF orbitals vanishing Hellmann Feynman forces have to be obtained at the energy minimum, the calculation yields values for the non H-atom at their crystallographic positions ranginging from 0.003 to 0.01 a.u.. This testifies to the excellent quality of the basis set. In the HF approach similar quality has only been reached for diatomics with Slater type orbitals.

Dipole Moments and Polarizabilities

In the last example the LDF approach has been used to calculate electrostatic properties from first principles. Calculated dipole moments shown in Table 8 and static polarizabilities shown in

Table 9 for a variety of molecules can be compared with experiment and calculations at the Hartree Fock and MP2 level.

Table 8: Dipole Moments [12] (Debye)

molecule	μ_{LDF}	μ_{exp}	μ_{HF}	μ_{MP2}
HF	1.78	1.80		
H_2O	1.73	1.85		
HCN	2.97	2.98		
NH_3	1.35	1.47		
H_2CO	2.32	2.33		
formamide	3.93	3.73		
imidazole	3.72	3.67	3.82	3.83
pyridine	2.29	2.19	2.67	2.35
p-nitroaniline	7.66			
adenine (A)	2.39	3.0	2.41	*2.1*
cytosin (C)	6.93	7.0	7.90	6.94
guanine (G)	7.26		7.80	*7.1*
thymine (T)	4.63	4.1	5.08	*4.3*
uracil	4.73	4.2	5.20	4.43
A-T	1.10			
G-C	7.00			

in italics: values extrapolated to MP2

LDF calculations with a numerical basis set including a single polarization function on the heavy atoms were found to provide reliable predictions for the dipole moments and static dipole polarizabilities for various molecules including nucleic acid bases. The values obtained correspond well with those calculated from more computationally demanding double zeta + polarization function MP2 calculations. Note that the MP2 values in italics are not obtained from a full MP2 calculation but rather by extrapolation. The true MP2 polarizabilities are in markedly better agreement with LDF than the extrapolated ones. It is not known at present why the LDF values for the polarizabilities of adenine, guanine and thymine do not agree better with experiment. The results for the base pairs suggest that the hydrogen bonds linking the two bases allow fractional charge transfer across the bonds.

Table 9: Polarizabilities [12] average (a.u.)

molecule	α_{LDF}	α_{exp}	α_{HF}	α_{MP2}
HF	5.8	5.6		
H_2O	10.0	11.7		
NH_3	14.4	15.2		
H_2CO	17.5	16.5		
C_2H_4	27.3	28.7		
formamide	28.2	27.5		
imidazole	48.8		44.8	46.9
pyridine	63.4	62.0	59.3	61.3
p-nitroaniline	107.7	105.9		
adenine (A)	95.8	88.4		*88.7*
cytosin (C)	79.6	69.5		76.7
guanine (G)	106.7	91.8		*94.9*
thymine (T)	85.2	75.8		*76.1*
uracil	71.9			68.1
A-T	186.7			
G-C	190.4			

in italics: values extrapolated to MP2

SUMMARY

The above examples illustrate some of the applications of local density functional calculations on large molecules and cluster models of chemical interest. A wide range of quantum chemical problems involving sizeable molecules are ready for study. In the coming years researchers will be able to use the LDF approach as a tool to calculate energy surface related and electrostatic properties reliably without prior assumptions. Such calculations will be of practical use by complementing and enhancing knowledge gained by experimental and other theoretical methods.

Acknowledgement: the author is indebted to Paul Jasien and George Fitzgerald for allowing him to quote their results prior to publication.

1 Kohn W, Sham LJ (1965) Phys Rev 136:B864
2 Delley B, Ellis DE, (1982) J Chem Phys 76:1949
3 Delley B, J Chem Phys in press
4 Versluis L, Ziegler T (1988) J Chem Phys 88:322
5 Chou SH, Freeman AJ, Grigoras S, Gentle TM, Delley B, Wimmer E (1986) J Am Chem Soc 109:1880 and (1988) J Chem Phys 89:5177
6 Ye Ling, Freeman AJ, Delley B (1989) Phys Rev B 39:10144
7 Delley B, (1989) Euro Phys Lett 10:347
8 Stevens J, Leung PC, Chou SH, Freeman AJ, Wimmer E, Delley B (1988) Cray Science and Engineering Symposium

9 Becker P, Cohen-Addad C, Delley B, Hirshfeld FL, Lehmann MS in (1986) Applied Quantum Chemistry 361, Reidel, Dordrecht Boston Lancaster Tokyo
10 Delley B, (1986) Chem Phys 110:329
11 Moeckly P, Schwarzenbach D, Bürgi HB, Hauser J, Delley B (1988) Acta Cryst B44:636
12 Jasien PG, Fitzgerald G submitted

Density Functional Calculations with Simulated Annealing – New Perspectives for Molecular Calculations

R. O. Jones and D. Hohl

Institut für Festkörperforschung, KFA Forschungszentrum Jülich, D-5170 Jülich

Abstract: The structure of a molecule or solid can be determined by calculating the total energy of the system of ions and electrons for all geometries. We review the problems inherent in such calculations, and show that the combination of density functional calculations with molecular dynamics techniques addresses the main difficulties. The method is applied to structural determinations in sulphur clusters S_n, where the ground state geometries are described very well. The method also gives interesting results in cases where there are structural changes involving large barriers (S_7O), and small energy differences with energy barriers on a thermal scale (isomers of Se_xS_y). As a final example, we discuss recent results on small phosphorus clusters, $P_n, n = 2, 8$.

1. Introduction

In principle it is easy to determine the ground state structures of molecules and solids. One chooses a structure, determines the total energy E of the system of electrons and ions, and then repeats this calculation for all possible geometries. The ground state structure is then that with the lowest energy. We shall see, however, that this method is impracticable in all but the smallest clusters. Not only is the calculation of the total energy difficult, but the number of local minima in the energy surface increases dramatically with increasing cluster size.

The combination of density functional (DF) methods, with a local density approximation for the exchange-correlation energy, and molecular dynamics (MD) with "simulated annealing" addresses both of these problems. After outlining the important features of the approach, we give examples to show that reliable geometries are found for small clusters of sulphur. Molecular dynamics also allows us to follow large changes in molecular structures, as we show for S_7O. For heterocyclic clusters of the form Se_xS_y, the method describes accurately the energy differences between different isomers. Finally, we describe the results of calculations for small phosphorus clusters, which have been the subject of considerable interest and controversy.

U. Harms (Ed.)
Supercomputer and Chemistry
© Springer-Verlag Berlin Heidelberg 1990

2. Determination of Structural Properties

2.1 Calculation of the total energy, E

A knowledge of the exact wave function, Ψ, of an interacting system of electrons and ions, would allow one to determine many quantities of interest, including the total energy

$$E = \langle \Psi | \widehat{\mathcal{H}} | \Psi \rangle / \langle \Psi | \Psi \rangle, \tag{1}$$

where $\widehat{\mathcal{H}}$ is the Hamiltonian of the system. Calculations of this type can be carried out for systems with very few electrons, the best-known example being the hydrogen atom. Even for the helium atom, however, it is not possible to determine the wave function and energy in closed form, and approximations for Ψ are unavoidable.

Approximate solutions for Ψ are usually based on the variational principle of Rayleigh and Ritz, i.e. if $|\Phi\rangle$ is an approximate wave function, then

$$\langle \Phi | \widehat{\mathcal{H}} | \Phi \rangle / \langle \Phi | \Phi \rangle \geq E_{GS}. \tag{2}$$

The simplest form of the many-particle wave function is that of Hartree (1928), who represented Φ as a product of single-electron functions:

$$|\Phi\rangle = \phi_1(\mathbf{r}_1)\phi_2(\mathbf{r}_2)...\phi_n(\mathbf{r}_n), \tag{3}$$

where the functions ϕ_i each satisfy a Schrödinger equation whose potential term is given by the mean field of the other electrons.

$$\left(-\frac{1}{2}\nabla^2 + V_{ext} + \varphi\right)\phi_i(\mathbf{r}_i) = \epsilon_i \phi_i(\mathbf{r}_i), \tag{4}$$

where V_{ext} is the external field of the nuclei, and the Coulomb potential φ is determined by solving Poisson's equation [we use atomic units, $e = m = \hbar = 1$]. The incorporation of Fermi statistics by replacing the product by a single (Slater) determinant leads to an additional potential term (the "exchange" potential), without changing the single-particle picture. The result (known as the "Hartree-Fock approximation") has been the basic method of atomic and molecular physics for the past 60 years.

A single determinant, however, is not generally sufficient to obtain a satisfactory total energy. In fact, Coulson[1] noted thirty years ago that "it is now perfectly clear that a single configuration wave function must inevitably lead to a poor energy". While a representation of the wave function as a linear combination of determinants ("configuration interaction", CI) should lead in principle to an exact wave function and energy, the numerical effort required increases explosively with increasing electron number. It remains difficult in practice to obtain accurate energies for systems with more than ~ 20 electrons.

Table 1. Number of minima in cluster interacting with a Lennard-Jones potential [Eq. (5)] (Hoare and McInnes, Ref. 2.)

N	LJ
6	2
7	4
8	8
9	18
10	57
11	145
12	366
13	988

2.2 Determination of the lowest-lying minima

These arguments apply to a *single* geometry, i.e. for a particular set of "internal coordinates" of a system. If we have N atoms in a cluster, the total number of internal coordinates is $3N - 6$, since we obtain the same energy if we rotate or translate the cluster. In a 10-atom cluster, there would be 24 independent interatomic degrees of freedom, so that the magnitude of the problem is obvious: (1) Finding the energy minimum by calculating the energy for all possible geometries is impracticable, since 10 calculations per coordinate would lead to 10^{24} calculations in a cluster where *one* seems to be too difficult! (2) Even if we could locate the closest minimum to a particular geometry, there is little prospect of success. We now show that the number of local minima in the energy surface can be very large.

Hoare and McInnes[2] have studied clusters where the atoms interact with each other with a simple, pairwise Lennard-Jones potential:

$$V_{Lennard-Jones} = 4\epsilon_0 \left[\left(\frac{\sigma}{r}\right)^{12} - \left(\frac{\sigma}{r}\right)^6 \right] \quad (5)$$

For small clusters it is possible to locate all the minima, and the results are shown for clusters of different sizes in Table 1. The number of minima increases rapidly with increasing N, with signs that the increase is exponential. In fact, for clusters of identical atoms interacting with a pairwise potential, Wille and Vennik[3] have shown that there is no known algorithm that grows with time as power of N and with which one can determine the ground state energy and structure. Such problems are classified as "$NP - hard$"[4] and are termed "intractable". It is quite sobering to discover that mathematicians view a problem of widespread interest to be insoluble in practice. Moreover, parametrized force laws, such as those used by Hoare and McInnes, are inherently semi-empirical, and it is difficult to know how the choice of parameters affects the final results. *No* pairwise force can stabilize the diamond structure, and there are many other examples where three- and four-body forces are essential.

This discussion shows that there are two distinct problems to be solved, with *multiplicative* difficulties. We now describe an approach that addresses (if not solves) both aspects. First, the density functional formalism provides a parameter-free, numerically efficient scheme for calculating the *total energy* of the system of ions and electrons. We avoid the problem of calculating the exact wave function of the interacting system of electrons and ions. Second, we introduce the strategy of simulated annealing as a procedure for determining low-lying energy minima, i.e. for finding solutions to the minimization problem that are *almost* optimal.

2.3 Energy calculations using density functional theory

The two basic theorems of the density functional formalism were derived by Hohenberg and Kohn.[5] They showed that:

(1) Ground state (GS) properties of a system of electrons and ions in an external field, V_{ext}, can be expressed as functionals of the electron density, $n(\mathbf{r})$, i.e. they are determined by a knowledge of the density *alone*. The total energy, E, is an example.

(2) $E[n]$ satisfies the variational principle $E[n] \geq E_{GS}$, and the density for which the equality holds is the ground state density, n_{GS}.

A simple and general proof of these assertions has been provided by Levy.[6]

The applicability of this scheme results from the observation by Kohn and Sham[7] that the minimization of $E[n]$ is simplified if we write:

$$E[n] = T_0[n] + \int d\mathbf{r}\, n(\mathbf{r})\left(V_{ext}(\mathbf{r}) + \frac{1}{2}\varphi(\mathbf{r})\right) + E_{xc}[n], \qquad (6)$$

where T_0 is the kinetic energy that a system with density n would have in the absence of electron-electron interactions, $\varphi(\mathbf{r})$ is the Coulomb potential, and E_{xc} is our definition of the exchange-correlation energy. The variational principle now yields

$$\frac{\delta E[n]}{\delta n(\mathbf{r})} = \frac{\delta T_0}{\delta n(\mathbf{r})} + V_{ext}(\mathbf{r}) + \varphi(\mathbf{r}) + \frac{\delta E_{xc}[n]}{\delta n(\mathbf{r})} = \mu, \qquad (7)$$

where μ is the Lagrange multiplier associated with the requirement of constant particle number.

If we now consider the corresponding equation for a system of *noninteracting* particles,

$$\frac{\delta E[n]}{\delta n(\mathbf{r})} = \frac{\delta T_0}{\delta n(\mathbf{r})} + V(\mathbf{r}) = \mu, \qquad (8)$$

we note that the mathematical problems (7) and (8) are *identical*, provided

$$V(\mathbf{r}) = V_{ext} + \Phi(\mathbf{r}) + \frac{\delta E_{xc}[n]}{\delta n(\mathbf{r})}. \qquad (9)$$

The solution of Eq. (8) is straightforward, requiring only that we solve the Schrödinger equation for noninteracting particles,

$$(-\frac{1}{2}\nabla^2 + V(\mathbf{r}))\,\psi_i(\mathbf{r}) = \epsilon_i \psi_i(\mathbf{r}),$$

$$n(\mathbf{r}) = \sum_{i=1}^{N} |\psi_i(\mathbf{r})|^2. \tag{10}$$

We have then reduced the problem of determining the total energy of a system of electrons and ions *exactly* to the solution of a single-particle equation of Hartree form (cf. Eq. 4). The solution of the equations leads to the energy and density of the lowest state, and all quantities derivable from them. It shares the main advantage of the HF method, namely the single-particle interpretation of the results. In contrast to the Hartree-Fock potential, however, the effective potential has a *local* dependence on the density.

All terms in the energy expression (6) are straightforward to evaluate, apart from the exchange-correlation energy, E_{xc}. This term is relatively small, at least for atoms and molecules, and there is an exact expression for it:[8]

$$E_{xc} = \frac{1}{2}\int d\mathbf{r}\, n(\mathbf{r}) \int d\mathbf{r}'\, \frac{1}{|\mathbf{r}-\mathbf{r}'|} n_{xc}(\mathbf{r}, \mathbf{r}'-\mathbf{r}), \tag{11}$$

where

$$n_{xc}(\mathbf{r}, \mathbf{r}'-\mathbf{r}) \equiv n(\mathbf{r}')\int_0^1 d\lambda\, \left(g(\mathbf{r},\mathbf{r}',\lambda) - 1\right). \tag{12}$$

E_{xc} can then be understood as the interaction between $n(\mathbf{r})$ and the exchange-correlation hole, $n_{xc}(\mathbf{r})$. The function $g(\mathbf{r},\mathbf{r}',\lambda)$ is the pair-correlation function of a system with density $n(\mathbf{r})$ and electron-electron interaction $\lambda\varphi$. The evaluation of such correlation functions generally requires a knowledge of the exact wave function, which is precisely one of the problems we should like to avoid. It is therefore essential to develop and test approximations for E_{xc}.

The most widely used approximation is the local spin density (LSD) approximation

$$E_{xc}^{LSD} = \int d\mathbf{r}\, n(\mathbf{r})\, \varepsilon_{xc}[n_\uparrow(\mathbf{r}), n_\downarrow(\mathbf{r})]. \tag{13}$$

Here $\varepsilon_{xc}[n_\uparrow, n_\downarrow]$ is the exchange and correlation energy per particle of a homogeneous, spin-polarized electron gas with spin-up and spin-down densities $n_\uparrow$ and $n_\downarrow$, respectively. There are several possible parametrizations: In the work described here we use that of Vosko et al.[9] This approximation is free of adjustable parameters, but its application to atoms, molecules and solids cannot be justified by small departures from homogeneity. A detailed discussion is given in the review article by Jones and Gunnarsson.[10]

There have been many DF applications to small molecules using this approximation,[10] including group VI A molecules.[11] Ozone (O_3)[12] and thiozone (S_3),[13] for example, are described very poorly by a single determinant wave function, and Hartree-Fock calculations give a ground state with the wrong symmetry. By contrast, the LD-approximation gives a good description of both the ground state geometry and the energy difference between the ground state and the lowest state with bond angle $\alpha = 60°$. In S_3, there are two states with quite different geometries that are almost degenerate, and this prediction has been confirmed by CI calculations.[14]

2.4 The molecular dynamics/ density functional approach

In systems where the ground state is unknown or there are many local minima, it is necessary to develop alternative methods for finding solutions that are near to optimal. Kirkpatrick et al.[15] noted the connection between statistical physics and the minimization of a function of many variables, and suggested "simulated annealing" based on a Monte Carlo sampling as a way of finding such solutions. It has been shown recently[16] that simulated annealing can lead to *nearly* optimal solutions of special NP-hard problems in polynomial average time.

Molecular dynamics (MD) provides an alternative to the Monte Carlo approach, and Car and Parrinello[17] showed that it could be combined with the density functional (DF) scheme for calculating total energies. The result is a method for calculating electronic properties that is free of adjustable parameters and makes no assumptions about ground state geometries. Finite temperature MD techniques allow an efficient sampling of the potential energy surface, and the DF method,[10] with the local spin density (LSD) approximation for the exchange-correlation energy, avoids the parametrization of the interatomic forces common in MD schemes.[18]

There are *two* minimization problems to be solved: The total energy must be minimized for each geometry, and we must find the geometry with the lowest energy. This is done conveniently by viewing E as a function of two interdependent sets of degrees of freedom, $\{\psi_i\}$ and $\{\mathbf{R}_I\}$, and using standard MD techniques.[17] The energy E is determined from the density via the orbitals ψ_i, $n(\mathbf{r},t) = \sum_i |\psi_i(\mathbf{r},t)|^2$ and the density functional expression (6, 13) for the total energy E. The artificial second order Newton's dynamics for the electronic degrees of freedom (with electron "masses" μ_i chosen accordingly) ensure that the electrons remain close to their ground state. If we define a "temperature" T by the mean classical kinetic energy of the atoms, the atomic configurations observed during the annealing process represent physical molecules at elevated temperatures.

3. Application to clusters of group VI A elements

The group VI A elements provide some of the best characterized of small atomic clusters. These elements are unique in that many allotropes comprise regular arrays of well-separated ring molecules, and X-ray structure analyses have been performed for $S_n, n = 6 - 8, 10 - 13, 18, 20$[19,20] and $Se_n, n = 6, 8.$[19,21] S_9 has been found as microcrystals, and S_{18} exists in two distinct forms. Several mixed crystals of the form $Se_n S_m$ and a range of sulphur oxides and ions are also known. The preparation of these clusters has been reviewed by Steudel.[20]

Sulphur and selenium have much in common and it is not surprising that mixed isomers coexist. In addition to the intrinsic interest in $Se_x S_y$ molecules, an understanding of their structures could provide insight into the complex structures of related liquid and amorphous materials, including Se itself. Mixed molecules have been studied in the vapour phase,[22,23] as liquids and as solid solutions.[24,25,26] The different species crystallize together, leading to structures with sulphur and selenium atoms distributed over the atomic sites.[27] The possible complexity is evident from the example of eight-membered rings, where 30 different crown-shaped isomers can occur.

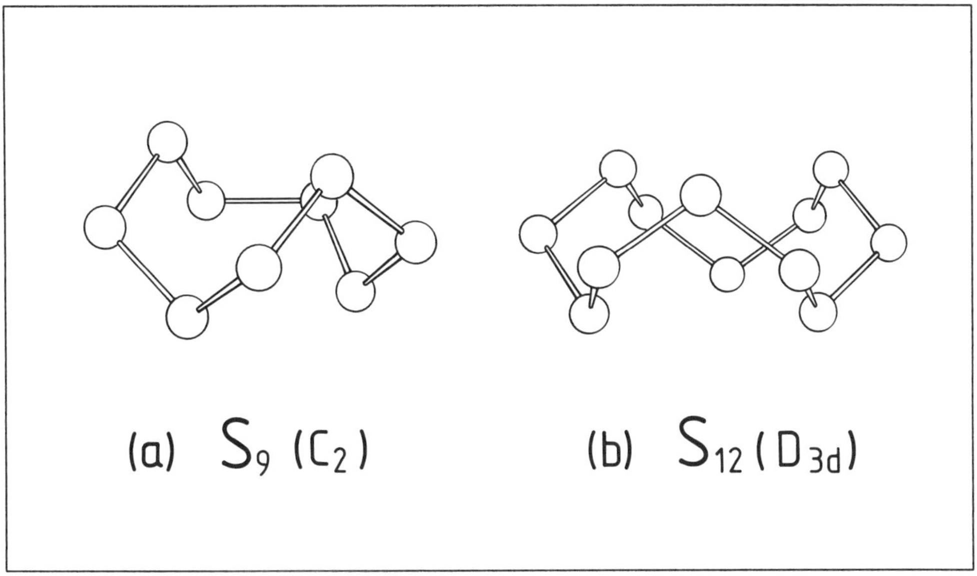

(a) S_9 (C_2)   (b) S_{12} (D_{3d})

Fig. 1. Structures of (a) S_9 and (b) S_{12}.

We have performed MD/DF calculations for some of these systems, using a large unit cell [constant volume 1000 Å³; fcc lattice constant 15.9 Å] with periodic boundary conditions. This guarantees a weak interaction between the clusters. The O, S and Se atoms are described by non-local pseudopotentials,[28] using pseudopotential components for $v_\ell(r), \ell = 0, 1, 2$. The energy cut-off for the plane wave expansion of the eigenfunctions, ψ_i, was taken to be 5.3 or 7 Hartree a.u., corresponding to $\sim$ 4000–6000 plane waves, except for S_7O, where a cut-off of 22.5 a.u. (36000 plane waves) was used. We use a single point $k = 0$ in the f.c.c. Brillouin zone. The "mass" of the electronic degrees of freedom was taken to be 300 a.u., and the time step for the simulations was $\Delta t = 1.7 \times 10^{-16}$ s (7.3×10^{-17} s for S_7O).

The results for sulphur clusters S_2 to S_{13}[29] show that it is possible to determine low-lying energy minima even if one starts from a geometry that is quite different. In S_{3-6} we started from almost linear chains and in S_{7-13} from nearly planar rings. After heating and cooling the system repeatedly, and ultimately cooling the molecule to $T = 0$, the final structures agreed well with experiment in all cases where X-ray data were available. As an example we show in Fig. 1 the structures of S_9 and S_{12}. The D_{3d} symmetry in S_{12} is reproduced very closely, and the structural parameters (bond length $d = 3.97$ a.u., bond angle $\alpha = 106°$, dihedral angle $\gamma = 88°$) are in good agreement with measured values (3.88 a.u., 106.2°, 87.2°, respectively).

Perhaps even more interesting is the application of the method to cases where the ground state structures are not yet known. In the case of S_9, for example, the crystallites that can be prepared are too small for X-ray diffraction analysis. We find a ground state isomer with C_2 symmetry [Fig. 1(b)] as well as several local minima with C_s symmetry with energies > 0.2 eV higher. The calculated ground state structure is consistent with measured Raman spectra, which indicate almost equal bond lengths and bond angles, and dihedral angles in the range $70 - 130°$.

3.2 Structural change in S_7O.

An interesting example of a structural optimization is provided by the S_7O molecule.[30] In the course of the simulation shown in Fig. 2 – 1200 time steps at $T = 2000$ K – the molecule changed from a stable local minimum with a ring structure similar to that in S_8 [Fig. 2(a)] to a structure with different topology, namely an oxygen atom outside the S_7 structure [Fig. 2(i)]. If we reduce the temperature of the system slowly, we find the geometry shown in Fig. 2(j). This agrees well with the experimental ground state.[30] It is remarkable that the entire structural change took place in only ~ 8000 time steps ($< 10^{-12}$ s), although the energy difference between ring and ground state structures is only ~ 0.2 eV and the energy barrier between them ~ 5 eV. It appears that the molecular dynamics technique is a particularly efficient generator of geometrical configurations.

3.3 Isomers of Se_2S_6, Se_6S_2.

There has been a considerable amount of work on S_nSe_{8-n} molecules. While earlier kinetic work led to conclusions that were to some extent conflicting, the situation has been improved in recent years by the use of Raman and NMR spectroscopies to study both these molecules and molten mixtures of Se and S.[31] Theoretical predictions of the relative stability of different isomers are difficult, since the interconversion of various Se_xS_y molecules and ring-chain equilibration processes involve the rupture and formation of $S - S$, $Se - Se$ and $Se - S$ bonds. Reactions involving the groups

$$-S-S- \quad + \quad -Se-Se- \quad \longrightarrow \quad 2 \; -Se-S- \qquad (14)$$

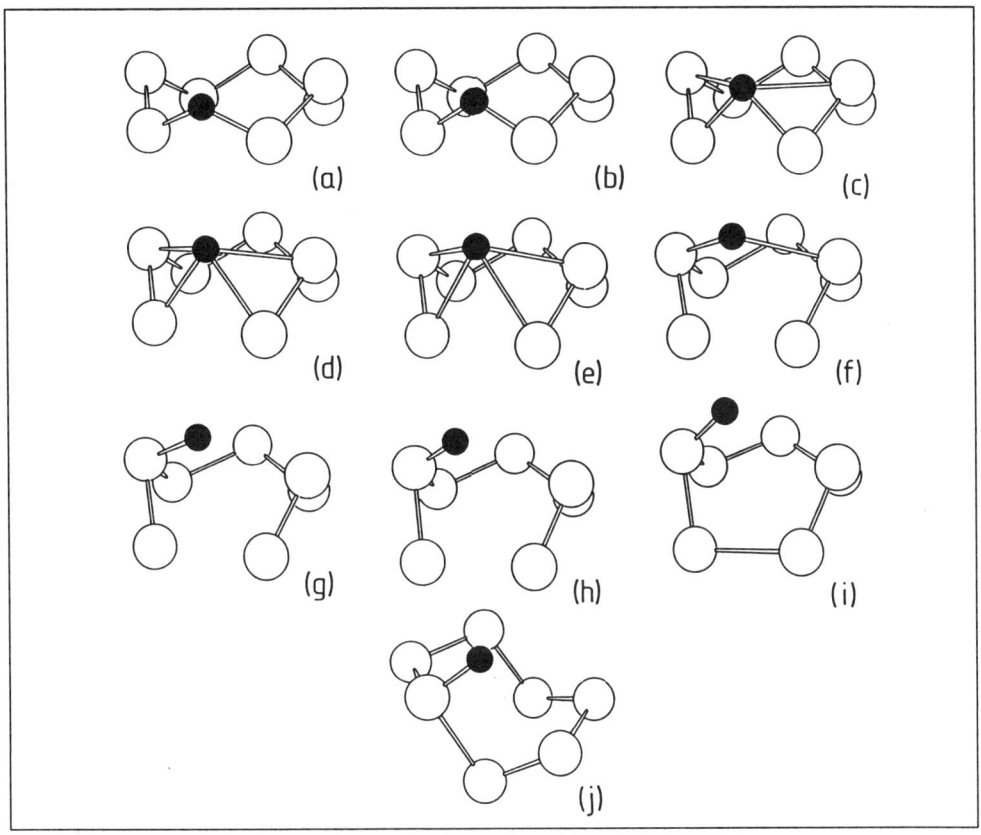

Fig. 2. Evolution of S_7O from ring (a) to ground state structure (j).

have been studied recently using Hartree-Fock techniques with a minimal Gaussian basis,[32] without reaching the accuracy needed for predicting the small energy differences involved. The reaction (14), for example, is predicted to be exothermic, in contrast to calorimetric measurements in the liquid.[33] For diatomic molecules in the gas phase,[34] the measured enthalpy change is $+7.9 \pm 6.7$ kJ mol^{-1} and the sum of the dissociation energies on the left side of (14) is greater than the sum on the right.

We have performed extensive calculations on seven- and eight-membered rings of the type Se_xS_y.[31] In each case we started with the ground state geometry of S_8 (D_{4d}), and determined the ground state geometry by alternating steepest descents techniques ($\Delta t = 1.0 \times 10^{-15}$ s, $\mu_i = 50000$ a.u.) with MD at $T = 200$ K ($\Delta t = 8.5 \times 10^{-17}$ s, $\mu_i = 300$ a.u.) In the simulated annealing runs, the average instantaneous temperature $\langle T \rangle$ – the mean kinetic energy of the ions – was kept constant within a 10° tolerance by rescaling the ionic velocities uniformly with a factor $\sqrt{\langle T \rangle / T}$.[35] Each structure determination required a total of about 4000 time steps. The electronic system was kept within $\ll 0.01$ eV of its ground state, so that the system of electrons and ions departs very little from the Born-

Oppenheimer (BO) surface. Within the framework of the LSD approximation and the basis described above, this approach allows us to determine the relative energies of the minima to better than 1×10^{-3} eV. Each time step required approximately 1 second of CPU time on a Cray X-MP/416 computer.

The cyclic structures possible in Se_2S_6 and Se_6S_2 are shown in Fig. 3 and the structural parameters for Se_2S_6 are given in Table 2. The structures show several remarkable features. In all eight structures, the three bond lengths ($S-S$, $S-Se$, $Se-Se$) are not only the same within 0.01 Å (2.08 Å, 2.23 Å, 2.35 Å), but the $S-S$ and $Se-Se$ bonds are the same as found in our previous calculations for the ring structures of S_8 and Se_8. All three are simply the sums of the corresponding atomic radii.

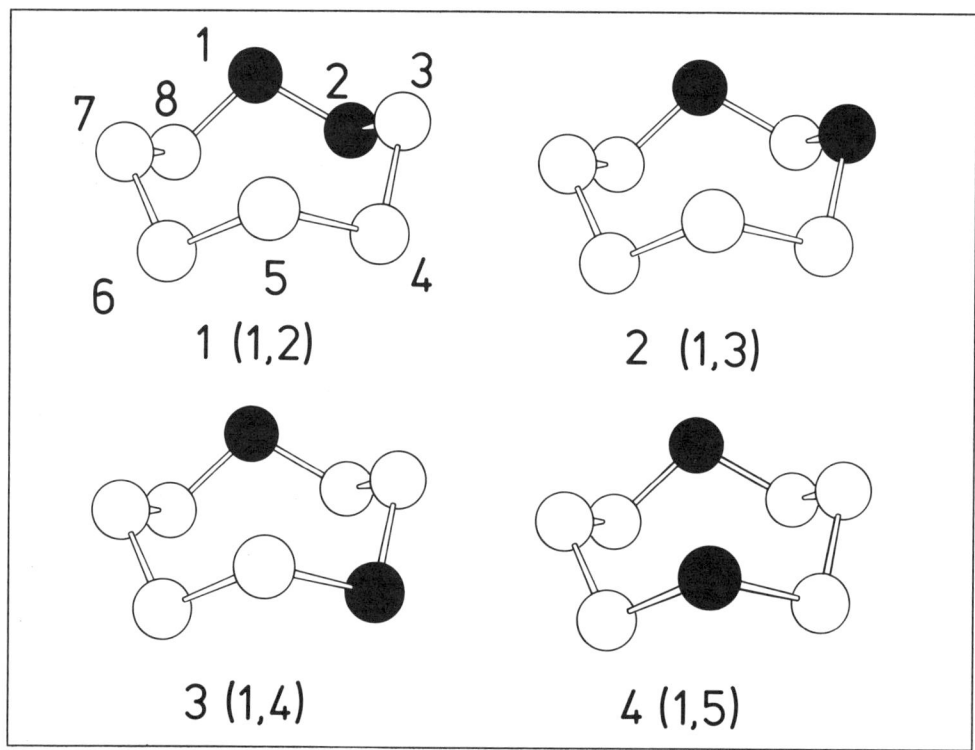

Fig. 3. Structures of Se_2S_6 (and Se_6S_2).

The energy calculations show that the (1,2) structures are the ground states in both Se_2S_6 and Se_6S_2. The energies of the other structures are *degenerate* to within 3 meV and lie 0.08 eV above the ground state in each case. We note here that the (1,2)-isomer is the most prominent detected in the NMR spectra of Se_2S_6 – there are smaller amounts of the other three – and the only one found in NMR spectra of Se_6S_2.[36] The calculations show some remarkable regularities, particularly in predicting ground states with adjacent minority atoms in each case, and we now show that the bonding trends can be understood in terms of a very

Table 2. Molecular parameters d, α and γ for all isomers of Se_2S_6 [see Fig. 3]. Bond lengths d_{ij} (between atoms i and j) in Å, bond angles α_i (at atom i) and dihedral (torsion) angles γ_{ij} (at bond ij) in degrees.

	(1,2) [C_2]	(1,3) [C_s]	(1,4) [C_2]	(1,5) [C_{2v}]
d_{12}	2.34	2.22	2.22	2.23
d_{23}	2.23	2.22	2.07	2.08
d_{34}	2.07	2.22	2.23	2.08
d_{45}	2.08	2.08	2.23	2.23
d_{56}	2.09	2.08	2.08	2.23
d_{67}	2.08	2.08	2.07	2.07
d_{78}	2.07	2.08	2.08	2.08
d_{81}	2.23	2.22	2.23	2.23
α_1	105.9	106.1	106.7	106.7
α_2	105.8	108.1	108.2	108.9
α_3	108.1	106.2	108.2	108.6
α_4	108.8	108.4	106.7	108.4
α_5	109.1	108.7	108.5	106.5
α_6	109.2	109.2	108.7	109.2
α_7	108.8	108.7	108.8	108.7
α_8	108.2	108.5	108.6	108.5
γ_{12}	96.8	98.0	98.3	99.1
γ_{23}	97.2	96.6	97.8	96.6
γ_{34}	97.6	96.1	96.3	98.2
γ_{45}	100.6	100.4	99.1	100.5
γ_{56}	102.9	102.1	101.1	99.0
γ_{67}	100.5	100.4	98.9	96.5
γ_{78}	97.5	99.0	99.1	97.7
γ_{81}	97.3	97.6	99.1	100.4

simple model.

The eight-membered rings are characterized by small deviations from the crown-shaped [D_{4d}] structures familiar from S_8 and Se_8. We have also seen that there are three bond lengths in the clusters, d_{S-S}, d_{S-Se} and d_{Se-Se}. The well-known relationship between bond length and bond strength[37] suggests that we associate with each bond type in these cyclic molecules corresponding contributions to the binding energy, E_{S-S}, E_{S-Se}, and E_{Se-Se}. In the case of the (1,2) structure in Se_2S_6, this leads to

$$E_{12} = E_{Se-Se} + 2E_{S-Se} + 5E_{S-S} \ . \tag{15}$$

For *all three* remaining structures, we find

$$E_{13,14,15} = 4E_{S-Se} + 4E_{S-S} \ . \tag{16}$$

From this simple argument we may then expect the energies of the last three structures to be equal, and the energy of the (1,2) structure to differ by an amount

$$\Delta E = E_{Se-Se} + E_{S-S} - 2E_{S-Se} . \qquad (17)$$

If we apply the same argument to the Se_6S_2 structures, Eq. (15-16) still apply, with S and Se interchanged. Again we find the (1,2) structure separated from the other three, and the energy difference is precisely that given in Eq. (17). This very simple argument explains why the energy orderings in Se_2S_6 and Se_6S_2 are the same, and is consistent with the most stable (1,2) isomers being separated from three structures with almost identical energies. The separation in the present calculations is approximately 0.08 eV (7.4 kJ mol^{-1}), which is consistent with the endothermicity of the reaction (14) and in reasonable agreement with the measured heats of reaction in the gas and liquid phases.

The results of the simple model agree with trends found in recent measurements, in particular with the relative abundance of structures with adjacent minority atoms. The symmetry between the energy ordering of the isomers of Se-rich and S-rich molecules is in direct contrast to earlier discussions of these systems.[22,38] This symmetry is perturbed, of course, by differences between the elements such as the size, but the essential features are not changed. The trend to "segregation" of minority components could also occur in the disordered liquid and amorphous states. It is important to note that such a segregation is a consequence, in this model, of the sign of ΔE in Eq. (17). Other systems with a small value of ΔE or with one of opposite sign would have a different behaviour.

4. Structure of phosphorus clusters, P_2 to P_8.

The final example we discuss here involves small phosphorus clusters. In elemental form, phosphorus shows a structural variety exceeded only by sulphur, and there have been many studies of the allotropes as well as amorphous forms.[19,39,40] Gas phase clusters have been of interest for many years. A problem of continuing interest has been the finding of only P_2, P_3, P_4 and (possibly) P_8 in the vapour phase.[41,42] There have been numerous speculations about the apparent absence of heavier clusters,[43] and several calculations indicating that the cubic form of P_8 is, in fact, less stable than two P_4-tetrahedra.[44,45] Recently, Martin[46] was able to detect mass spectroscopically P_n clusters up to $n = 24$. In spite of this, almost nothing is known about the structure of clusters with $n > 4$.

We have performed an extended series of MD/DF calculations on phosphorus clusters up to P_8.[47] The calculated and experimental geometries agree well in those cases (P_2, P_4) where the latter are known. In the larger clusters there were several interesting and unexpected results: (1) Although the tetrahedral structure is energetically favoured in P_4, there is a large "basin of attraction" for a D_{2d} "roof" structure, i.e. this structure is the closest minimum for a large region of configuration space. (2) The "roof" structure is a prominent feature in the low-lying isomers in P_5 to P_8. The calculated ground states in P_5, P_6, and P_7 have a P_4-roof with an additional one, two and three atoms, respectively. In the last case, the trimer is approximately equilateral.

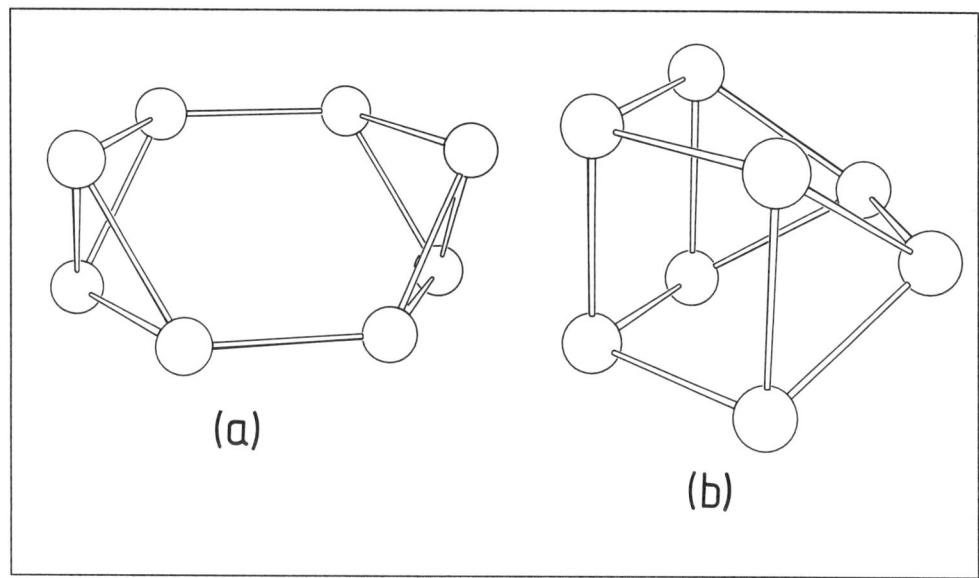

Fig. 4. Two structures of P_8.

In P_8, we have found two structures that are energetically more stable than the cubic structure and correspond to local minima in the energy surface. They are shown in Fig. 4. The structure in Fig. 4(a) shows again the preference for the "roof" structure, with two such units being bonded together. The ground state we found is shown in Fig. 4(b). It can most easily be understood as a (distorted) cube with one bond rotated through 90°. This "wedge" or "cradle" structure is found as a structural unit in violet (monoclinic, Hittorf) phosphorus.[48] There is a pronounced analogy between the structures of the P_8-isomers and those of the corresponding isoelectronic hydrocarbons C_8H_8. The cubic form of the latter (cubane) has been prepared by Eaton and Cole,[49] and can be converted catalytically to the wedge shaped form cuneane.[50] The third possible isomer of C_8H_8 [analogous to Fig. 4(b)] was estimated to have a strain energy intermediate between the other two. This agrees with the energetic ordering we have found in the P_8 isomers.[47]

We believe that the results for phosphorus clusters demonstrate clearly the usefulness of the simulated annealing approach. The calculations reproduce known structures very well, so that we may have confidence in the predictions for those molecules where the structures are presently unknown. The predictions show very plausible trends, even though several of the structures were quite unexpected.

5. Concluding remarks

The combination of molecular dynamics and density functional methods has given a very satisfactory description of the structures of small sulphur and selenium clusters, and has shown (in the case of S_7O) that the system of electrons and ions can undergo large structural changes involving energy barriers in excess of 5 eV. In S_7O, the final structure (an O atom outside an S_7-ring) is in very good agreement with the structure determined by X-ray diffraction. In the heterocyclic ring molecules Se_xS_y we have studied the opposite extreme — structures with very small energy differences and energy barriers that are small on a thermal scale. The results are very encouraging. In the case of phosphorus clusters, we predict structures with plausible trends and with an interesting analogy to isoelectronic hydrocarbons. It should be emphasized again, however, that there are many local minima in the energy surface for low-symmetry clusters of this size, and *no* method can guarantee finding the absolute energy minimum.

Several factors are important in enabling the MD simulated annealing procedure to locate prominent energy minima, in spite of the complexity of the energy surfaces and large energy barriers. (1) The supply of kinetic energy via the external heat bath is essential to overcome potential energy barriers that are large on a thermal scale (> 5 eV at $\langle T \rangle = 2000$ K in S_7O). This process is effective unless the energy surface is so steep that the kinetic energy loss cannot be compensated at the imposed $\langle T \rangle$. (2) The kinetic energy can be partitioned among the individual atoms of the molecule in a non-uniform way, so that single atoms in a small cluster can acquire much of the available kinetic energy. (3) The trajectories generated by the MD equations of motion appear to be very successful in locating reaction pathways to low-lying minima. (4) Our experience with group VI A clusters has shown that the lowest-lying minima are significantly more extended than higher ones. While this is true to some extent in the phosphorus clusters, we found a large "basin of attraction" for the D_{2d} structure of P_4, even though the energy lies more than 2 eV above the tetrahedral ground state.

We thank the German Supercomputer Centre (Höchstleistungsrechenzentrum, HLRZ) for a grant of computer time on the Cray Y-MP 832 in the Kernforschungsanlage Jülich.

References

1. Coulson C (1960) Rev Mod Phys **32** 170
2. Hoare MR, McInnes JA (1983) Adv Phys **32** 791
3. Wille LT, Vennik J (1985) J Phys A **18**, L419, L1113

4. Garey MR, Johnson DS (1979) *Computers and Intractability: A Guide to the Theory of NP-Completeness* Freeman, San Francisco

5. Hohenberg P, Kohn W (1964) Phys Rev **136** B864

6. Levy M (1979) Proc Nat Acad Sci USA **76** 6062

7. Kohn W, Sham LJ (1965) Phys Rev **140** A1133

8. Gunnarsson O, Lundqvist BI (1976) Phys Rev **13** 4274. See also Harris J, Jones RO (1974) J Phys F **4** 1170

9. Vosko S, Wilk L, Nusair M (1980) Can J Phys **58** 1200

10. For a recent survey of this formalism and some of its applications, see Jones RO, Gunnarsson O (1989) Rev Mod Phys **61** 689

11. For a discussion of O_3, SO_2, SOS and S_3, see Jones RO (1987) Adv Chem Phys **67** 413 and references therein

12. Jones RO (1985) J Chem Phys **82** 325

13. Jones RO (1986) J Chem Phys **84** 318

14. Rice JE, Amos RD, Handy NC, Lee TJ, Schaefer HF III (1986) J Chem Phys **85** 963

15. Kirkpatrick S, Gelatt CD, Vecchi MP (1983) Science **220** 671

16. Sasaki GH, Hajek B (1988) J Assoc Comput Machin **35** 387

17. Car R, Parrinello M (1985) Phys Rev Lett **55** 2471

18. Stillinger F, Weber TA, LaViolette RA (1986) J Chem Phys **85**, 6460

19. Donohue J (1974) *The Structures of the Elements* Wiley, New York, Chapters 8 [group V A] and 9 [group VI A]

20. Steudel R (1984) In: Müller A, Krebs B (eds) *Studies in Inorganic Chemistry, Vol. 5*, Elsevier, Amsterdam, p. 3

21. Steudel R, Strauss, EM (1984) Adv Inorg Chem Radiochem **28**, 135

22. Cooper R, Culka JV (1967) J Inorg Nucl Chem **29** 1217

23. Schmidt M, Wilhelm E (1970) Z Naturforsch **25** B 1348

24. Steudel R, Strauss EM (1987) In: *The Chemistry of Inorganic Homo- and Heterocycles, Vol. 2*, Academic, London, p. 769

25. Steudel R, Laitinen R (1982) Topics in Current Chemistry, **102**, 177

26. Bitterer H (ed) (1984) *Selenium: Gmelin Handbuch der Anorganischen Chemie*, 8. Aufl., Ergänzungsband B2. Springer, Berlin Heidelberg New York

27. Laitinen R, Niinistö L, Steudel R (1979) Acta Chem Scand A **33** 737

28. Bachelet GB, Hamann DR, Schlüter M (1982) Phys Rev B **26**, 4199
29. Hohl D, Jones RO, Car R, Parrinello M (1988) J Chem Phys **89** 6823
30. Hohl D, Jones RO, Car R, Parrinello M (1989) J Am Chem Soc **111** 825
31. Jones RO, Hohl D (1990) J Am Chem Soc (in press)
32. Laitinen R, Pakkanen T (1983) J Mol Struct (Theochem) **91** 337; (1985) J Mol Struct (Theochem) **124** 293
33. Maekawa T, Yokokawa T, Niwa K (1973) Bull Chem Soc Japan **46** 761
34. Drowart J, Smoes S (1977) J Chem Soc, Faraday Trans II **73**, 1755
35. Woodcock LV (1971) Chem Phys Lett **10** 257
36. Laitinen RS, Pakkanen, TP (1987) Inorg Chem **26** 2598
37. For a discussion of the relationship between core size, valence eigenfunctions and bond strengths, see Harris J, Jones RO (1979) Phys Rev A **19** 1813
38. Eysel HH, Sunder S (1979) Inorg Chem **79** 2626
39. Corbridge DEC (1974) *The Structural Chemistry of Phosphorus*, Elsevier, Amsterdam
40. Corbridge DEC (1985) *Phosphorus. An Outline of its Chemistry, Biochemistry and Technology* (Third Edition). Elsevier, Amsterdam
41. Kerwin L (1954) Can J Phys **32** 757
42. Carette JD, Kerwin L (1961) Can J Phys **39** 1300
43. Bock H, Müller H (1984) Inorg Chem **23** 4365
44. Raghavachari K, Haddon RC, Binkley JS (1985) Chem Phys Lett **122** 219
45. Ahlrichs R, Brode S, Ehrhardt C (1985) J Am Chem Soc **107** 7260
46. Martin TP (1986) Z Phys D **3** 221
47. Jones RO, Hohl D (submitted to J Chem Phys)
48. Thurn H, Krebs H (1969) Acta Cryst B **25** 125.
49. Eaton PE, Cole TW Jr (1964) J Am Chem Soc **86** 962, 3157
50. Cassar L, Eaton PE, Halpern J (1970) J Am Chem Soc **92** 6366

Authors' Index

Barnickel, G., E. MERCK, Frankfurter Str. 250, 6100 Darmstadt

Burkhardt A., Max-Planck-Institut für Festkörperforschung,
7000 Stuttgart 80

Delley B., Paul Scherrer Institut, c/o Laboratories RCA,
Badenerstr. 569, CH - 8048 Zürich

Gentzsch, W., FH Regensburg, Pruefeningerstr. 58, 8400 Regensburg

Grüner, M., CESYS, Gesellschaft für Angewandte Mikroelektronik,
8520 Erlangen

Häußler, G., CESYS, Gesellschaft für Angewandte Mikroelektronik,
8520 Erlangen

Hohl D., Institut für Festkörperforschung, KFA Forschungszentrum
Jülich, 5170 Jülich

Jones R.O., Institut für Festkörperforschung, KFA Forschungszentrum Jülich, 5170 Jülich

Köppen, H., Boehringer Ingelheim KG., Med. Chem. Dept.,
6507 Ingelheim

Meier U., Ruhr Universität Bochum, Lehrstuhl für Theoretische
Chemie, Universitätsstr. 150, 4630 Bochum

Streitz, S., PARACOM GmbH, Jülicher Str. 338, 5100 Aachen

v. Schnering H.G., Max-Planck-Institut für Festkörperforschung,
7000 Stuttgart 80

Vogelsang R., SUPRENUM GmbH, Hohe Str. 73, 5300 Bonn 1

Weber, H.P., SANDOZ AG, Preclinical Research, CH-4002 Basel

Wedig, U., Max-Planck-Institut für Festkörperforschung,
7000 Stuttgart 80